The Inequality of Man

H. J. Eysenck

The Inequality of Man

Temple Smith · London

First published in Great Britain 1973
by Maurice Temple Smith Ltd
37 Great Russell Street, London WC1
© 1973 Hans J. Eysenck
Reprinted 1974
ISBN 0 8511 7050 1
Printed in Great Britain by
Ebenezer Baylis & Son Limited
The Trinity Press, Worcester, and London

Contents

Tables

Figures

To Gary, Connie, Kevin and Darrin
– in the hope that genetic regression to the mean
has not dealt too harshly with them!

'To sequester out of the world into *Atlantick* and *Eutopian* politics, which never can be drawn into use, will not mend our condition; but to ordain wisely as in this world of evil, in the midst whereof God hath placed us unavoidably.'

JOHN MILTON

CHAPTER I

Equality and Individuality

This book has a very simple purpose. There is much talk about various political objectives, such as the improvement of education, the elimination of social classes, and the creation of communities in which human equality is achieved to a much greater extent than is true of our present societies. There are also objections to these arguments, in terms of necessary loss of liberty, or the incompatibility of equality with human nature. Proponents and objectors usually base their arguments on preconceived ideas; both postulate the kind of human nature which suits their purpose. In this book I have asked the simple question—what does science have to say on the subject? What knowledge have psychology, genetics, and physiology contributed in the past few decades which would enable us to regard these issues dispassionately, and come to a conclusion less strongly based on preconceptions? More generally, do these sciences have anything to say that we should heed in making decisions on social policy? I believe very strongly that in many areas—education, penology, social mobility, to take just some examples almost at random— there does already exist sufficient knowledge to make it unwise for society to disregard these contributions of science to social welfare. I shall give several examples to show how society, by making erroneous assumptions, has in fact achieved the opposite effects to those intended; these errors could have been predicted, and correct action taken, had factual knowledge been available to those responsible for making the policy decisions in question.

Quite generally, my argument is based on a very simple premise. Man is a biological as well as a social organism; evolutionary doctrine teaches us that our brain, our autonomic system, and our whole bodily structure have evolved through millions of years in response to a very hostile environment. What has singled us out from all other animals has been the unique development of the neocortex, that great mass of white and grey matter inside our skull which enables us to function intelligently, to adjust rationally, and to solve problems successfully. Our entire culture is based on this fragile basis; without our intellect we would be wiped out in very little time by animals much better equipped in many ways to survive in a world created by 'nature red in tooth and claw'. Evolution proceeds by selection, and selection is based on the existence of genetically determined individual differences.

It seems very likely, since evolution proceeds by selection, that innate differences among human beings extend to such complex traits and abilities as are manifested in intelligence and personality, in mental illness and in criminality. Much is now known about these questions, although we are of course still far from reaching any final conclusions: it seemed worth while to try and put the sum and substance of our knowledge into a form which would be intelligible to educated readers without any special knowledge of psychology or genetics.

The first part of the book consists of material which demonstrates the differences between people, with respect to intelligence, personality, predisposition to mental disorder, to crime, and to psychopathic behaviour. Some of the points made there are not new, but there has in fact been such a flood of novel material, outstanding in quality and highly sophisticated in design, that even books written just a few years ago are now decidedly out of date. It is only last year (1972) for instance, that research material has been published on the children of psychopathic and criminal parents, put out to adoption at an early age: when these children were compared with children of normal parents, put out to adoption at an equally early age, and matched with them on a whole array of important variables, it was found that these former children showed very significant tendencies to behave like their biological parents, rather than like their adoptive parents. Such studies complement the many studies of twins which had previously been the only evidence to show that criminal conduct is to a marked extent determined by genetic causes.

In the field of intelligence testing, too, there has been much new material. I have been particularly concerned not only to incorporate this material into the book but also to deal with certain perennial objections which are voiced against IQ testing, and the measurement of intelligence as a whole. These objections are based largely on a misunderstanding—a misunderstanding of how science in general defines and investigates certain concepts which it finds useful, and a misunderstanding of how psychologists in fact proceed in studying intelligence. There are, in addition, certain recent developments, such as the growth of interest in 'creativity', or in the concepts advocated by Piaget, which are said to make traditional IQ testing old-fashioned and out-of-date. I shall show that this is not so, and I shall attempt to discuss the general paradigm of modern intelligence testing theory and technology briefly but I hope intelligibly. While these discussions are in one sense marginal to my main purpose which is to demonstrate the social consequences of the dependence of IQ on genetic causes, they are nevertheless necessary; if there were any weaknesses in the demonstration of this dependence, or in the very concept of 'intelligence' as measured by IQ tests, then the whole consequent argument would clearly fall to the ground.

Later in the book I turn to the topic which I personally find most interesting, and which unfortunately has not been at all widely discussed, namely the impact which our growing knowledge of innate differences must have on society as a whole. From the very moment that I was introduced to the literature on intelligence as a young student, it seemed to me that our whole thinking about society, about politics, and about education must be powerfully influenced by this new body of knowledge; I have looked in vain for any recognition of this fact, or indeed any sign that a serious debate was about to begin on these issues. When, later on, I came into contact with the evidence concerning the influence of heredity on personality, on criminality and on mental disorder, and had a chance to work myself in these fields, it seemed to me even more certain that our whole way of looking at social problems solely in terms of simple environmental manipulation was wrong, and that a great deal of new thinking was required to incorporate this new knowledge into our general philosophy. I have tried to set out some arguments in this book as to how this might be done, but I am far more concerned to get a discussion going than to attempt to give any final answers. There are issues here which are clearly of the utmost import-ance; it would be a tragedy if they were swept under the carpet because current fashion is opposed to thinking in terms of genetic predispositions, and recognizes only the most simple and elementary form of environmental manipulation.

If our problems were as simple as many people seem to believe, they would not have plagued us for thousands of years; nor would they continue to produce almost in-superable obstacles to a solution now. Any positive sugges-tions I have to make are quite likely to be wrong, but they should surely be submitted to informed criticism: neither studied neglect nor active violence are a proper answer.

In this book I have attempted to review the literature on these subjects, trying to give a generally understand-able account of what is beginning to be recognized as a reasonably accurate model of human behaviour. Un-fortunately much of this literature is of necessity mathematical: our model must be quantitative if it is to be properly testable. Readers who are not satisfied with my attempts to render a non-mathematical account will find a more adequate and detailed one in my book on *The Measurement of Intelligence*; this also contains a full set of references to the material here only briefly alluded to. In the references I mention a few standard texts of recent origin, to enable the reader to consult orthodox opinion on issues where he might feel inclined to probe a little more deeply, without necessarily wishing to go to the original sources. For those who do not like the conclusions reached, there is really no alternative but to become professionally interested in the subject, to read up the literature on the theory and testing of intelligence and personality, on behavioural genetics and on the various other topics touched on: rejection of scientific conclusions because of conflicting preconceptions is not a rational method of dealing with the problem.

It is dangerous to quote objections made by some authority, many years ago, of findings then published and interpreted in certain ways; more recent work may easily have rendered these objections untenable. This is particularly true in the genetic field, where very important advances have been made literally within the past two or three years.

A simple example may show how strong preconceived ideas can be. It is a curious fact that most people nowadays almost instinctively adopt an environmentalist explanation and strongly resist a genetic one.

Consider the cries of young babies who want food, or who are surprised (novel stimulus), or who are frustrated, or who greet their mother after an absence. Behaviourists would say, and most people would take this for granted, that babies learn the sounds to make under these distinct conditions. The notion that the specific noises appropriate to these four conditions might be innate, very much as the young bird's specific song is largely innate, would be received with incredulity.

Derek Ricks recorded the cries of many babies, emitted when they were exposed to the four conditions described above; he then presented the cries of groups of six babies to sets of parents, including in each group the baby belonging to these particular parents. Parents were asked (1) to identify their own baby, which they thought would be very easy, and (2) to identify the specific request of the babies, which they thought would be impossible. In actual fact, few parents succeeded in identifying their own babies, but they almost all succeeded in identifying the conditions which caused the babies to emit their cries. There clearly is an innate pattern of cry production which ensures that babies emit certain types of cries in certain types of situation; mothers recognize these cries, even when (as in some of the experiments) the baby comes from a different national or racial background.

Autistic children, it is interesting to note, were found to lack this genetic pattern; their cries were unique and clearly learned in the 'environmental' fashion—mothers recognized their own autistic babies, but did not recognize the meaning of the cries of autistic babies not their own. (They did recognize the cries of normal babies!) This has a bearing on the environmentalist theories of autism, as opposed to genetic theories, but it would take us too far to go into this question now; let us merely note that even the earliest cries emitted by normal babies are strongly based on a genetic foundation, and that current beliefs in the simple environmentalist hypothesis of their causation are quite mistaken.

The past history of the battle between those who believe in the fundamental equality of all human beings (in the sense of equality in mental and physical characteristics of social importance, not in the sense of equal worth in the eyes of God, or of equal rights before the law) can be traced in three publications which carried titles similar to the one chosen for this volume. The first of these is Rousseau's *Discours sur l'origine de l'inégalité parmi les hommes*, which was published in 1754; the ideas contained therein were repeated eight years later in his much better known *Du contrat social*. Although every notion of fundamental importance in these books comes from the works of Hobbes and Locke, nevertheless it is to Rousseau that most egalitarians in the next century or two turned for the most vivid expression of their ideas.

In 1890 T. H. Huxley undertook to answer Rousseau from the point of view of a biologist; his article appeared in the pages of the well-known journal, the *Nineteenth Century*. 'On the Natural Inequality of Men' presents the biological evidence available to early followers of Darwin; in the nature of things there is no mention of the Mendelian principles of genetics, and consequently the arguments presented are less strong than they might have been had Huxley been possessed of our present-day knowledge. But even so, his is a shrewdly aimed counter-blast, and many of his arguments still deserve attention. We will consider them in a moment.

The third author to use the title was another biologist, very different in many ways from Huxley. J. B. S. Haldane wrote 'The Inequality of Man' as an essay, and later published it, together with other essays, in a book which was published in 1932, under the same title. Where Huxley was a conservative (with a small 'c'), Haldane was a leading Communist; where Huxley was ignorant of the principles of genetics, Haldane was a world-renowned authority on the subject. It is interesting to note that in spite of these far-reaching differences, both men arrive at very similar conclusions, as we shall see. Haldane put his respect for scientific facts before his political opinions: when the two came into conflict, politics had to take second place.

Haldane proclaimed the facts of human inequality in intelligence and personality on the basis of such studies as were familiar to him; we now know far more than he did, of course, by virtue of the great advances which have been made since his time, but in essence what he wrote then would still be true today. A great scientist sniffs out the truth even from partial and often insufficient evidence. Haldane recognized a fact when he encountered it, and did not attempt to accommodate the facts to his own preferences. He, too, of course, was a great writer; there are few popularizers in science who could compete with Haldane, and his essay is still a joy to read.

To return to the debate, as Huxley points out:

> Undoubtedly, Rousseau's extremely attractive and widely read writings did a great deal to give a colour of rationality to those principles of '89 which, even after the lapse of a century, are considered by a good many people to be the Magna Charta of the human race. 'Liberty, Equality, and Fraternity' is still the war-cry of those, and they are many, who think, with Rousseau, that human suffering must needs be the consequence of the artificial arrangements of society and can all be alleviated or removed by political changes.

Huxley goes on to say that:

> I have very long entertained the conviction that the revived Rousseauism of our day is working sad mischief, leading astray those who have not the time, even when they possess the ability, to go to the root of the superficially plausible doctrines which are disseminated among them.

Rousseau, in fact, begins his *Discours* by making a distinction between two kinds of inequality, very much as do Huxley and Haldane later on:

> the one which I term *natural*, or *physical*, because it is established by nature, and which consists in the differences of age, health, bodily strength, and intellectual or spiritual qualities; the other, which may be called *moral*, or *political*, because it depends on a sort of convention, and is established, or at least

authorised, by the consent of mankind. This last inequality consists in the different privileges which some enjoy, to the prejudice of others, as being richer, more honoured, more powerful than they, or by making themselves obeyed by others.

So far, so good: note that Rousseau puts intelligence among the natural or physical differences; unlike modern egalitarians he does not consider differences in intelligence to be due to political and other social causes.

Having begun so reasonably, however, Rousseau soon abandons reason. As Huxley puts it so clearly:

> Of course the question readily suggests itself: Before drawing this sharp line of demarcation between natural and political inequality, might it not be as well to inquire whether they are not intimately connected, in such a manner that the latter is essentially a consequence of the former? This question is indeed put by Rousseau himself. And, as the only answer he has to give is a piece of silly and insincere rhetoric about it being a question fit only for slaves to discuss in presence of their masters, we may fairly conclude that he knew well enough he dare not grapple with it. The only safe course for him was to go by on the other side and as far as the breadth of the road would permit; and, in the rest of his writings, to play fast and loose with the two senses of inequality, as convenience might dictate.

This question, translated into modern terms, will indeed occupy us to quite a considerable extent; rephrased, it might read: Can it be that the class structure of modern society is essentially a function of the innately differing intellectual and other qualities of the people making up these classes? Is the status which a person enjoys in society, as a function of his profession, determined by the same innate qualities? If we wish for complete, or at least greater equality than exists at present, how can we set about achieving this aim, or does it go completely counter to biological reality?

How does Haldane interpret the discoveries of modern psychology and genetics in the context of the questions raised by Rousseau and Huxley? He begins by noting that 'in the present age the admirable institution of universal suffrage is . . . supported by the curious dogma of the equality of man'. He goes on to say that 'human inequality springs from two sources, nature and nurture. . . . Some inequality due to differences of environment is inevitable, if only because of the facts of geography. But in its grosser forms it means an immense waste of human possibilities and every progressive State aims at equality of opportunity.' This statement is immediately followed by a strong emphasis on the fact that 'men are not born equal'. Haldane quotes in support such strange bedfellows as Napoleon ('La carrière ouverte aux talents') and Jesus ('who converted the word "talent" from the name for a sum of money to an expression for inborn human ability, of which he clearly recognized the existence'). Oddly enough, Haldane does not quote the communist adage 'To each according to his need, from each according to his ability'; this too surely recognizes innate human differences in both needs and abilities. (The saying refers to the condition obtaining once the communist state has been achieved, so that any remaining differences could not be due to environmental factors supposedly eliminated by communism.)

Haldane next discusses some of the evidence for genetic causes of inequality in intelligence, quoting some of the earliest studies of foster children (whose IQs resemble those of their natural mothers, rather than those of their foster mothers) and of identical twins reared in separation (where the separated twins nevertheless show very similar IQs). He uses these findings to discredit two extreme views: 'Today extreme eugenists proclaim that environment has very little influence, extreme behaviourists that nothing else matters.' Both views, he suggests,

are mistaken, and he believes that 'the progress of biology in the next century will lead to a recognition of the
innate inequality of man'. He points out that 'this is
today most obviously visible in the United States, where
educational opportunities are more widespread than
elsewhere. Universal education leads, not to equality,
but to inequality based on real differences of talent.'

After giving a short account of the findings on intelligence testing, and of the nature of intelligence, Haldane
makes some prophecies which will sound curious to his
political fellow-travellers:

> It should be possible by the time a child is about
> seven to arrive at a fair idea of its capacities, and
> children will be sorted out accordingly. Today we
> often have special schools for mentally deficient
> children, and occasionally for very able ones. This
> system will, of course, be greatly extended. When
> children of all grades of ability are combined in one
> class, the intelligent merely learn to be lazy while the
> stupid are hopelessly discouraged. And the attempt
> to remedy this defect by placing children of widely
> different ages in the same class is also a failure.

So much for comprehensive schools in the socialist
millennium. Altogether, 'in a scientifically ordered
society innate human diversity would be accepted as a
natural phenomenon like the weather, predictable to a
considerable extent, but very difficult to control.'
Haldane does not believe

> that a recognition of the inequality of man would be
> a blow to democracy (or rather to representative
> government based on universal suffrage) . . . It is,
> of course, irrational that each man's vote should
> possess equal value. But the alternatives so far tried
> or suggested are still less rational. They usually take
> the form of increasing the political power of those
> who are wealthy enough to be able to influence
> politics already.

And he concludes with another look into the future:

> Some day it may be possible to devise a scientific
> method of assessing the voting power of individuals.
> One can be fairly certain that that day is more than
> a century ahead. In the remote future mankind may
> be divided into castes like Hindus or termites. But
> today the recognition of innate inequality should
> lead not to less, but to greater, equality of oppor-
> tunity.

What Haldane, like Huxley before him, is trying to
point out is that equality before the law, equality of
opportunity, and equality as a citizen, are not dependent
on *identity* of genetic endowment; *these are human rights, of
universal validity, which are independent of biological and other
scientific findings*. Indeed, it would be extremely dangerous
to argue (as some modern theorists do) that people are
universally entitled to these equal rights *because* they are
genetically equal; if science should prove that such
genetic equality was a myth, rather than a fact, then the
whole case for equality as a human right would fall to
the ground. As will be shown in this book, science does
indeed show that genetic equality is a myth; this does
not in any way lead to an argument favouring inequality
before the law, inequality of opportunity, or inequality
of political rights.

Recognition of inequality, of human diversity, only
refers to specific traits, abilities, behaviour patterns; it
does not imply general superiority or inferiority. Extroverts are different from introverts; they are neither inferior nor superior. Traits and abilities which are considered useful in the context of our present civilization
may not be so considered in the context of some future
(or past) civilization. Judgments of 'superior' or 'inferior' are mostly culturally determined statements based
on premises which have no scientific status. A possible
exception to this rule is presented by the numerous
physical and mental diseases which have a strong genetic

basis, but it is still possible to argue that genius may often be allied with madness, as the poet suggested. Intelligence may seem to be an obvious exception, but even here society, in order to survive, needs diversity; a nation of Einsteins or Newtons would not be viable!

There is no question whatever that men are created unequal, in the sense that their genes contain the determinants of unequal appearance and development; it remains the task of research to seek out the precise mode of interaction between heredity and environment, and to determine the limits which, at any given moment of time, the one sets to the influence of the other. Egalitarianism is an aspiration, not a fact; it is a meaningful question to ask to what extent it is compatible with biological knowledge, and, should we find that it is not, to what extent it is compatible with liberty—because in order to impose equality on a non-egalitarian biological substratum, we may have to curtail the liberty of the individuals concerned and try to force them into a uniform mould.

We must come to grips with the factual problem of the degree to which heredity shapes our abilities, what these abilities are, and to what extent they can be measured. And we must also try to deal with the much more elusive problem of the social effects of these factual discoveries; do they really lead to greater equality, rather than to less? How can we best use the knowledge acquired for the benefit of our children—not some of our children, but all of them? Is it reasonable to refuse to take genetic facts into account in devising our educational system, on *a priori* grounds of egalitarian sentiment? Are there ways of reducing genetic inequalities by environmental manipulation? Clearly the factual and the political problems are closely connected and intertwined. The former, of course, can receive a scientific answer to which a considerable degree of certainty attaches; the latter every reader must answer for himself.

It is unfortunate, but perhaps inevitable, that the

assumed close relation between behaviour-genetic fact and social-political attitude has produced highly polemical and unproductive discussions between adherents of extreme positions on either side. These discussions between 'environmentalists' and 'hereditarians' have of course a respectable antiquity; philosophers will remember the nativism of Kant and the *tabula rasa* of Locke, to give but one example. It is now mostly in psychology that this debate continues, and it does so in many different guises. Thus traditionally the study of perception emphasized learning and environmental factors, until the Gestalt school stressed maturation and genetic factors. MacDougall, in the field of social behaviour, attributed major importance to instincts, which of course had a strong genetic basis; his later critics decried the very existence of instincts, and postulated instead social learning. In both fields we are now beyond the simple-minded swing of the pendulum; we now have accumulated sufficient indisputable facts to know that both environmental and genetic factors interact to produce perception and social behaviour as we know it. Hubel and Wiesel have isolated specific neural elements which subserve specific types of perception, and the ethologists have demonstrated beyond question 'instinctive' modes of behaviour in many different animals. These discoveries not only reconcile extremist views, but they also transcend them; what the ethologists find is not really very close to MacDougall's instincts, and what Hubel and Wiesel find is not really quite what the Gestaltists anticipated.

In these fields, then, the pendulum has ceased to make the wild and exaggerated swings which are so characteristic of the recent past. In the fields of intelligence and personality, unfortunately, we are still in the middle of one of these 'swings', and a proper understanding of the material here discussed is not really possible without some knowledge of the developments which led up to the present position. We may usefully begin with the position as it was around the time of the First World War, and shortly after; a longer historical perspective is not required for our purposes. At this time, there was a close connection between social science and what passed for genetics; it was widely believed that mental illness, feeblemindedness, criminality and other deviations from normality resulted from abnormalities of the brain and central nervous system, and, being physiological in origin, were determined largely by hereditary causes. The evidence on which these beliefs were based was extremely poor by modern standards; it would be correct to say that none of it would be acceptable to workers in any of these fields nowadays. The fact that it was widely accepted may be connected with the fact that at that time the *Zeitgeist* in America (where most of this work was being done) was pervasively racist and nativistic.

American achievement was credited to the genius of the Anglo-Saxon race, and Madison Grant's book, *The Passing of the Great Race*, in 1916 constitutes a kind of apotheosis of this crudely pro-Nordic and anti-Semitic, anti-Negro attitude; it would not be correct to mention this monstrosity were it not for the fact that it was widely acclaimed in what we would regard as academic and scientific circles. Eugenics, based on this very crude hereditarianism, was widely adopted even in government circles, and laws were enacted for the sterilization of criminals, the insane, and the feeble-minded; in addition, Congress enacted immigration laws restricting the influx into the USA of groups considered 'racially inferior'. The kind of teaching that resulted in these excesses is exemplified by a speech delivered by C. B. Davenport, the nation's most influential human geneticist, who declared that social reform was futile and that 'the only way to secure innate capacity is by breeding it'. William MacDougall called for the replacement of democracy by a caste system based upon biological capacity, with legal restrictions upon breeding by the lower castes, and upon intermarriage between the castes.

Gradually the excesses of the hereditarian approach produced a strong opposition, which derived much support from the violent reaction to Nazi racialism and cruelty experienced all over the Western world. It became dogma to assert that an explanatory sequence that began with the organism inevitably ended with predatory ethics. There also emerged a political connection; Pastore in 1949 studied twenty-four participants in the nature-nurture controversy and found that of the 'hereditarians' only L. B. Terman (the originator of the famous 'gifted children' follow-up study in California, about which we shall have more to say later on) could be regarded as a liberal, while among the 'environmentalists' J. B. Watson, the originator of behaviourism, was the only conservative. Similarly, it has been found that stress on heredity is usually associated with having been educated at one of the prestige 'ivy league' universities, while stress on environment is more frequently found in students educated at state universities. These facts are often quoted in support of the view that one side or the other has adopted its scientific views because of bias generated by its political and social beliefs; but other explanations are possible. Perhaps some at least of the writers in question have adopted their political views as a consequence of being exposed to the scientific evidence? The evidence is lacking to decide between alternative explanations and in any case these are *argumenta ad hominem* which should not be used in scientific discussion; the facts are the only things that matter, and on those grounds there is no doubt that the hereditarians were largely in the wrong in what they asserted. Even on the few points where they were right, they were right by accident rather than on the basis of proper, scientifically admissible evidence.

However, it is an unfortunate fact of scientific controversy that the holder of one extremist view is not the best exponent for criticizing another extremist view; the pendulum tends to swing too far the other way and we end up in a position only slightly better than the one we started out in. Extremist hereditarianism was replaced by extremist environmentalism; socially this may be more acceptable, but scientifically it has proved no more closely related to the experimentally established facts. Sociology, criminology, psychiatry and education have based themselves on the shifting quicksands of a doctrinal environmentalism which may occasionally pay lip-service to the possible influence of genetic factors, but which in practice disregards the contributions of such factors completely and almost contemptuously. Experiments are designed so as to make it impossible to sort out the influence of heredity; causal explanations of any findings are sought exclusively in environmental terms, without even an acknowledgement of the possibility that heredity might have produced the same results: criticisms of genetic theories are made without any acquaintance with modern methods and results. Quite often research results are accepted because they are in accordance with the spirit of the times, although the methodology used in the experiment in question is dubious, the statistics faulty, and the conclusions unsupported by the evidence.

Facts, of course, are often slippery things; it is not always obvious just what are facts, and what artefacts. Several examples are given in this book of alleged facts which turn out on inspection to be nothing of the kind. It often needs an experienced eye to discover the error in an otherwise excellent study.

Consider for instance a large-scale study recently carried out in America on the relation between cancer of the womb and circumcision—the hypothesis being that wives of circumcised husbands are less likely to contract this disease, because of greater hygiene practised by circumcised men. Many thousands of women were investigated, and asked about their husbands' penile status; hundreds of hours were spent on the computer rendering down the data into digestible statistical fodder. When far-reaching conclusions had been drawn from the results, a psychologist asked quite innocently whether many of these women actually knew whether their husbands had in fact been circumcised! After much ridicule had been heaped upon his head a special investigation was done to answer this question, and the answer turned out to be—no! Quite a few women had only the haziest notion of what circumcision meant—some thought it meant wholesale amputation. Many women had never seen their husbands in the nude, and had no idea whether they had or had not been circumcised. The 'facts' of this investigation had been based on very fragile foundations indeed.

Modern human geneticists certainly refuse to view the world as a contest between heredity and environment. To quote J. M. Thoday (1965):   .

> Every character is both genetic and environmental in origin... Genotype determines the potentialities of an organism. Environment determines which or how much of these potentialities shall be realized during development. The doctrine of fixed abilities is nonsense. . . . When asking about the genetic factors or the environmental factors that influence intelligence, or any other characteristic, we are asking about the causes of variety in a population and . . . unless the populations have been artificially produced by special breeding programmes designed to eliminate genetic variety, the causes of variance are *always* both genetic and environmental and the variance can be partitioned into three components, genetic variance, environmental variance and variance arising from genotype-environment interaction.

Genetics has come a long way since the days of Davenport; so has psychology. We now know much more than we did then about intelligence, and about personality.

A proper reassessment is now possible of the entrenched positions of the hereditarianism of the 1920s and the environmentalism of the 1950s and 1960s; neither emerges with any great distinction. The only tenable view today (and indeed for many years previously) is one of *interactionism*, that is a view which seeks for both hereditary and environmental causes for individual differences in intelligence and personality, and which also takes into account the interaction between these variables.

We might end this chapter by stressing the importance of man's individuality and of his differing needs. Perhaps this diversity can be best illustrated by an example.

It is well known that men and women have somewhat divergent attitudes to many sexual questions; it is equally clear that within a given sex there are still tremendous differences between the most 'feminine' and the most 'masculine'—using these terms as purely descriptive of the attitudes held, and not as synonymous with homosexuality or lesbianism. Table 1 gives a list of questions

Sex without love ('impersonal sex') is highly unsatisfactory	Agree	*Disagree*
Conditions have to be just right to get me excited sexually	Yes	*No*
Sometimes it has been a problem to control my sex feelings	*Yes*	No
I do not need to respect a woman (man), or love her (him), in order to enjoy petting and/or intercourse with her (him)	*Yes*	No
It doesn't take much to get me excited sexually	*True*	False
I think about sex almost every day	*Yes*	No
The thought of a sex orgy is disgusting to me	Yes	*No*
I like to look at sexy pictures	*Yes*	No
Seeing a person nude doesn't interest me	True	*False*
I believe in taking my pleasures where I find them	*Yes*	No

If I had the chance to see people making love without being seen, I would take it	*Yes*	No
Pornographic writings should be freely allowed to be published	*Yes*	No
Prostitution should be legally permitted	*Yes*	No
There should be no censorship, on sexual grounds, of plays and films	*Agree*	Disagree
Absolute faithfulness to one partner throughout life is nearly as silly as celibacy	*Yes*	No
I would enjoy watching my usual sex partner having intercourse with someone else	*Yes*	No
Sex is more exciting with a stranger	*Yes*	No
To me, few things are more important than sex	*Yes*	No
Group sex appeals to me	*Yes*	No
The thought of an illicit relationship excites me	*Yes*	No
The idea of 'wife swapping' is extremely distasteful to me	Yes	*No*
I can take sex and I can leave it alone	Yes	*No*
Some forms of love-making are disgusting to me	Yes	*No*
If you were invited to see a 'blue' film, you would	*Accept*	Refuse
If you were offered a highly pornographic book, you would	*Accept*	Refuse
If you were invited to take part in an orgy, you would	*Accept*	Refuse

Table 1. *Typical questions from the masculinity-femininity scale of sex attitudes. In this questionnaire, the answers more frequently given by males have been set in italic*

asked of some four hundred men and an equal number of women; on each of these questions there were large differences in the proportions of the two sexes answering 'Yes' and 'No'. The typical male answer has been set in italics in each case, to make it clear how this test of masculinity-femininity is scored. Figure 1 shows the distri-

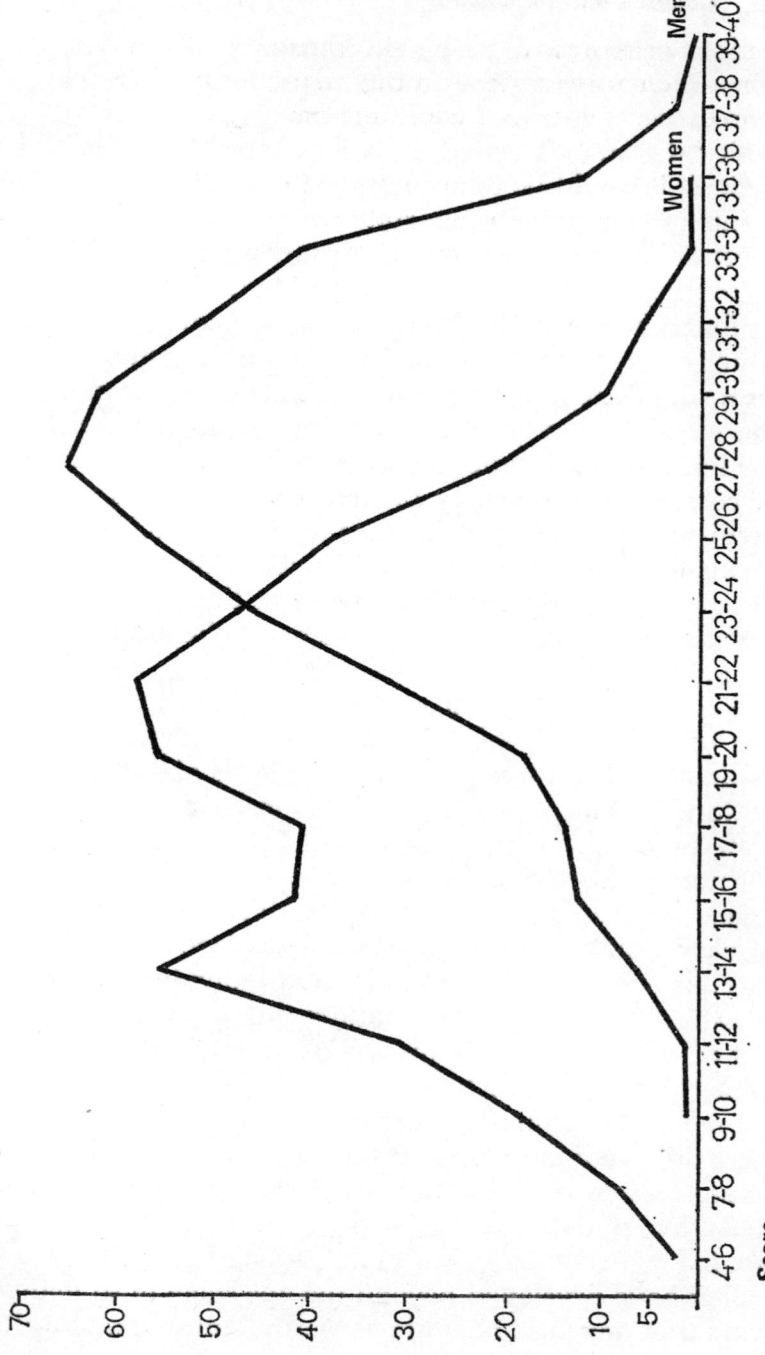

Number of Subjects:

Score

Figure I *Scores of men and women on masculinity-femininity sex attitudes inventory*

butions of scores made by the members of our sample when each answer given in the 'masculinity' direction is scored one point, and each answer given in the 'femininity' direction is scored no points. It will be clear how difficult it would be to accommodate all these divergent points of view within one state-decreed ethic, and how constraining and emotionally disturbing such enforced uniformity would be. Similar individual differences can be found in nearly all those aspects of communal living which make up one's personal life; the less the state interferes with these aspects of a person's conduct, the greater the effective liberty which that person enjoys (within the limits of not interfering with other people's liberties, of course). The diversity of man extends to his sexual predilections and preferences, as we can see in Figure 1, as much as to his IQ, or his extraversion; individual differences are all-pervasive. The more this fact is recognized by governments, the less likely are they to legislate for uniformity.

Some people feel that one or the other of these two sets of attitudes is 'right', the other 'wrong', in some meaningful way; usually the argument is about what is 'natural', what 'unnatural'. The masculine point of view would seem to be that marriage and monogamy are unnatural, and that all sorts of ways of satisfying one's sexual impulses are equally natural. The feminine point of view would seem to assert that 'impersonal sex' and 'perversions' are unnatural, and that individualized love is natural. Both appeal to 'nature', but such appeals do not get us very far. Shrimps and doves are indeed monogamous; on the other hand, rats and squids are not. Are we more like shrimps or squids? It is only necessary to raise the question to see that the whole argument is nonsense. Because baboons and lions live in groups with one male dominating a harem of dependent females, it does not follow that this is natural for human beings. Human beings have developed beyond the ape stage in so many ways that any argument about 'natural' and 'unnatural' is simply irrelevant, in so far as it brings in comparisons with lower animals. What we can say is that men and women are not 'equal' in the sense of having identical physical make-up and identical attitudes based on identical hormonal secretions; they are equal in the sense that the value judgments implicit in their attitudes have an equal right to be seriously considered. Similarly, the great individual differences within each sex which become so apparent in our Figure 1 depend on genetic factors to a marked extent, and cannot be dismissed as evidence of 'unnaturalness'. Whatever exists must *eo ipso* be natural—else how could it exist at all? The diversity of man is as apparent in his sexual and other personal behaviour as it is in his abilities, his criminal and psychiatric peculiarities, or his general personality, and rules and laws which do not take this fact into account are doomed to run into trouble—and cause untold misery to thousands of men and women in the process.

A recognition of this essential human diversity, based as it is on genetic factors at present beyond our control, would save us untold misery, and release our energies from a vain struggle to confine other people's behaviour within chains of our own devising. It is very difficult to think of someone else's views as being just as sane, just as reasonable, just as likely to promote happiness as our own, when his views conflict with ours. If there were some agreed premises from which we could derive our views through a process of logical argument, then we could hope to convince our opponents of their essential wrongheadedness, and demonstrate at least their lack of logic. But alas, there are no such agreed premises in most of these fields; to go back to our attitudes to sex, these are due in part to inherited differences in hormonal secretions, in part to a history of conditioning received during our early years. To argue about either is clearly futile. Argument will not change your inherited hormonal pattern, nor will it do much to affect your past history of reinforcement. Religious people believe that

2

their religion furnishes them with a firm basis from which agreed premises can be constructed, but adherents of other religions, or of none, are not bound by these in any way. Evolution teaches us that such genetic diversity, far from being a nuisance, is a blessing which cannot be overestimated; adjustment to changing circumstances is made possible by the existence of all those genes which may at any particular time seem either useless or an active obstacle to adjustment. For this reason tolerance towards other views, other patterns of living, other modes of adjustment, is not just a desirable aspect of democratic living; it is the essence of preserving our human capacity for change and adjustment to altered circumstances. The inequality of man needs as a counterpoise tolerance for individual differences; it also requires, from the political point of view, a better understanding of the causes of such differences. We are learning to ask the right questions; perhaps in due course we shall come up with the right answers.

There will, of course, be many refutations of the points made in this book, usually, if past history is a good guide, by people not familiar with all the facts and theories in question. There may be a faint resemblance to the days when relativity theory was condemned by the Russians, for being bourgeois, and by the Nazis, for being Jewish. Einstein made the classic reply when a book was published in Germany in which one hundred Nazi professors condemned his theory. 'Were I wrong,' he said, '*one* professor would have been quite enough.'

CHAPTER 2

What do IQ tests really measure?

The notion that IQ tests measure intelligence, as their name suggests, used to be accepted very widely, and their wide use in school, in industry and in the armed services of many countries made them invaluable instruments and tools for many social purposes. However, in recent years a large number of criticisms have been heard, and the picture has become decidedly blurred. Critics have complained that the very notion of 'intelligence' is mistaken, and that instead we should be speaking of many different intellectual abilities; they have in fact suggested that if 'intelligence is what intelligence tests measure', then the ability in question is as useless as prowess at solving crossword puzzles. Other critics have complained about the apparent arbitrariness of making up tests and scoring systems; they have suggested that these tests are constructed by white, middle-class psychologists for white, middle-class children—with the implication that the tests are unfair to working-class children, or to coloured children, or in fact to all 'deprived' groups. Race and social class are often cited as factors that make impossible the measurement of 'innate ability', and that make environmental influences predominant in our type of society.

Many critics, including many psychologists, have made fun of the difficulties which experts have had in agreeing on a definition of 'intelligence': there are many such definitions, ranging from 'ability to profit from experience' to 'speed of intellectual functioning', or from 'ability to reason' to 'capacity for learning', but none has been universally accepted. A few knowledgeable critics have pointed out that even the mathematical tools often used by psychologists working in this area to resolve theoretical conflicts (such as factor analysis) have failed to do so, and have given rise to disputations of their own, such as those relating to the quarrel between the followers of Spearman who favour a single, general factor of intelligence, that is, an ability used in all mental tasks, and the followers of Thurstone who prefer to postulate a number of primary abilities such as verbal ability, numerical ability or spatial ability, which are used in particular mental tasks. There is no doubt that these and many other criticisms—children can be coached on IQ tests, different tests do not give identical scores for the same children, some tests are speeded, others not—have created a climate in which the great contributions of IQ testing to psychological theory and social practice are being overlooked, and intolerance is being fostered against all 'mental measurement'. The true position is complex, and requires detailed discussion.

IQ tests were designed to measure intelligence; it is not unreasonable to ask if they in fact do so. Most people expect a simple answer; they probably feel that it should be easy to correlate IQ scores with a proper criterion. If that correlation,* derived from a proper sample of a given population, is high and positive, then they would be prepared to accept the proposition. This is one of the methods of studying the validity of a test recognized by psychologists, the so-called 'predictive' method: it is based on the notion that there is a generally agreed criterion for whatever it is we wish to measure, and that the validity of the test depends on how far it enables us to predict a person's score on that criterion. (The term 'validity' is somewhat technical; it simply means that the test measures what it is supposed to measure. The related term 'reliability' simply means

* Correlation is the degree to which statistical variables vary together. It is measured by the correlation coefficient which has value from zero (no correlation) to -1 or to $+1$, perfect negative or positive correlation.

that the test measures the same thing on separate occasions. To be useful, a test usually has to have both high reliability and high validity; in fact, it could hardly have high validity if it had low reliability.)

Unfortunately, we have no such agreed criterion for IQ—if we did, we would probably not need IQ tests anyway! Several criteria have of course been suggested, but they are all very partial and unsatisfactory. Teachers' ratings of a child's intelligence are obviously valuable, but they are far from perfect; in addition, different teachers often disagree about a particular child. School success has been suggested, but extremely intelligent persons (as shown by their later careers) often do badly at school—Winston Churchill and Charles Darwin are only two examples that spring to mind. Money-earning capacity used to be considered as another criterion, but few people would make that suggestion seriously in this day and age. Overall life success might be thought of as another criterion, but such factors as luck, illness, or just being born with the proverbial silver spoon in one's mouth obviously make such a criterion much less valuable than it would ideally be. The truth is that while we would expect intelligence to *correlate* reasonably highly with all these 'criteria', we cannot accept any of them as constituting a scientifically acceptable criterion: thus we are deprived of 'predictive validity' as our main support in proving that IQ tests really measure intelligence.

A second test of validity, *concurrent validity*, must also be eliminated. This simply consists in showing that the proposed measure correlates with another measure which has already been validated. Many IQ tests were thus 'validated' by being correlated with the original Binet-Simon test, the first-ever intelligence test which appeared in the early years of this century. But of course this does not tell us anything about the validity of the original Binet-Simon test; if we have reason to doubt our stipulated criterion, then the fact that our own test correlates well with it loses its value as evidence.

We could of course cut our coat according to our cloth and abandon the claim that our tests measure some abstract entity called 'intelligence' and simply say that we are measuring something like 'scholastic ability'. If we did that, then correlations between our tests and scholastic achievement would constitute predictive validity, and the fact that a given test correlated with another one that had been shown to have such predictive validity would constitute concurrent validity. There is no doubt that IQ tests do predict scholastic achievement with considerable accuracy, both at school and even at college and university. These correlations show certain interesting features: students with low IQ scores hardly ever do well, but while students with high IQ scores usually do well, a surprising number do rather poorly. Scholastic ability is not enough; clearly students also need motivation, persistence, and other personality attributes, as well as certain outside conditions, such as family support. The correlation between scholastic ability and achievement is much higher at school than at university; this is almost certainly due to the fact that there is a restriction of range of talent as you move up the educational ladder—at university, only the ablest have survived, and it is much more difficult to make distinctions among that élite. We will not discuss this possible way out here because we intend to claim that IQ tests do in some scientifically meaningful way measure 'intelligence', and hence this method of solving our problem does not seem necessary.

A third method of validation for a test is that of *content validity*. This is established by showing that the test items are a sample of a universe in which the investigator is interested. Here again the difficulty of defining intelligence makes it difficult for us to make use of this method; if we cannot define the universe of test items, then we cannot demonstrate that our test is made up of a fair sample of items from this hypothetical universe. There are of course ways out of this difficulty. British education

authorities have renamed the IQ tests used in pupil
selection 'tests of verbal reasoning', and it is probably
easier to think of a universe of items which would be
universally recognized as measuring up to this description. However, renaming of this kind seems a coward's
way out, although many people have adopted this
method. In any case, the solution this might present to
our problem is probably purely semantic and of no great
scientific interest.

We are left with two further alternatives. One of these,
while widely favoured by behaviourist psychologists,
will be dismissed very briefly. This is the method of
operational definition, a term introduced into the philosophy
of science by the physicist W. P. Bridgman in 1938. In
his book *The Logic of Modern Physics* he undertook a careful examination of the use of various concepts in current
physical theory, particularly those representing theoretical concepts. The cure he recommended for the
many abuses he discovered was the scrupulous recognition of the operations carried out by the experimenter in
the course of his observations and measurements. To
define a theoretical variable is to specify the experimental operations carried out for the purpose of its
measurement. Unfortunately, as C. L. Hull (1943) has
pointed out:

> his emphasis upon the operations which are the
> means whereby the observations and measurements
> in question become possible has led many psycholo-
> gists to mistake the means for the end. The point . . .
> to be emphasized is that while observations must be
> considered in the context of the operations which
> make them possible, the central factor in the situation
> is *what is observed* . . . An emphasis on operations which
> ignores the central importance of the dependent ob-
> servations completely misses the virtue of what is
> coming to be known as operationism.

It would be in good agreement with the letter of

operationism to define intelligence as 'that which intelligence tests measure', and this definition is sometimes
given by psychologists tired of eternal argument; nevertheless it would be against the spirit of Bridgman's
thesis. We would still be required to study carefully
what is observed, that is the nature of the problems included in the IQ tests, the methods adopted by subjects
in their attempts to solve these problems, and the meaning of the observed scores in the wider context of prediction. Operationism is a good doctrine, but it is not
enough for a proper understanding.

This leaves us with but one method of validating our
tests and concepts, namely a method labelled 'construct
validation' by Cronbach and Meehl (1955). 'Construct
validity must be investigated whenever no criterion or
universe of content is accepted as entirely adequate to
define the quality to be measured.' The method to be
used, in essence, consists of starting with some concept
(like 'intelligence' or 'temperature') for which there
does not exist any proper criterion; the nature of this
concept will specify various predictions which can be
tested, and which serve to validate (or invalidate) the
concept in its original form. If, as is most likely to happen,
some but not all of the predictions turn out as expected,
the original concept will have to be changed, and a new
round of tests carried out to validate the new concept
('intelligence$_2$' or 'temperature$_2$'), and so on until we end
up with a concept which for the time being is regarded
as adequate ('intelligence$_n$' or 'temperature$_n$').

Even a very successful concept may in due course have
to be given up (for instance gravitation) because some
facts are found to be absolutely irreconcilable with
crucial elements of the concept, and because a new concept (for instance space-time continuum and field
theory) are found to be more adequate. At any given
time the theory surrounding a concept specifies an infinite number of possible experiments which could be
carried out to test the adequacy of the concept; only a

very limited number of these can be tested empirically.
Thus we can never prove a theory to be correct; we can
only prove it to be incorrect. Even this is not easy: the
history of science shows that every far-ranging theory
has been beset by anomalies at all stages of its existence,
without causing those active in the given field to abandon the theory. The ideas of 'proof' or 'simple disproof'
as the basis of scientific advance have been abandoned,
as Thomas Kuhn and Karl Popper have made clear
(1970) (in *Criticism and the Growth of Knowledge*, edited by
Lakatos and Musgrave). Nevertheless, this hypothetico-
deductive approach has proved useful in the 'hard'
sciences, and there seems no reason why it should not
be equally useful in psychology.

As Cronbach and Meehl point out 'scientifically
speaking, to "make clear what something *is*" means to
set forth the laws in which it occurs'. These laws may be
deterministic or statistical, and at least some of these
laws must involve observable phenomena. A good example of the sort of thing that happens in science is the
development of the construct 'temperature'. To begin
with we have a very subjective state of immediate experience of 'hot' and 'cold' and 'lukewarm'; this does
not provide a proper criterion for scientific measurement
because of its obvious unreliability. Put your right
hand in a basin full of hot water, your left in a basin full
of cold water, and then put both hands in a basin full
of water of an intermediate temperature—it will feel cold
to the right hand, hot to the left! A room that is stiflingly
hot to an Englishman feels cold to an American, and so
on. Nevertheless, we would distrust any measure of
temperature that actually contradicted subjective experience, just as we would distrust any measure of 'intelligence' that disagreed completely with teachers'
ratings, or with examination results in school subjects.

We next find that the expansion of a column of air,
raising or lowering the level of water in the same glass
container agrees quite well with subjective estimates;

this provides us with a thermoscope, but with one that has many limitations. We have no means of calibrating it, and worse still, we have forgotten to close the top so that the measurement of temperature is contaminated by barometric pressure. Fludd, Santorio and their followers in the seventeenth century provided an arbitrary scale, but not until Pascal's famous account of his elaborate barometric experiments on the Puy-de-Dôme did the latter defect become apparent. As Pascal said, 'from (this experiment) there follow many consequences such as . . . the lack of certainty that is in the thermometer for indicating the degrees of heat (contrary to common sentiment). Its water sometimes rises when the heat increases, and sometimes falls when the heat diminishes, even though the thermometer has remained in the same place.'

Once the thermometer had been sealed, and once the imperfections of the air-water method had been recognized, the search began for a rational way of calibrating the instrument, that is, for some way of constructing a scale which would have a replicable bottom and top point. The long and often hilarious history of this search is well described in Middleton's *A History of the Thermometer*. It used to be thought, for instance, that all caves have identical temperatures, so that thousands of thermometers were dragged into all sorts of underground caves to be calibrated! In spite of the fact that many famous physicists took part in this work, progress was slow and the measurement of temperature remained haphazard for a long time. Even now, of course, the ordinary thermometer has many disadvantages. It is not very accurate; it only measures a minute portion of the temperature scale (from a little below freezing to a little above boiling) so that other means have to be sought for measuring nearer absolute zero, or above the boiling point for metals; it is subject to change with age. These points made about the measurement of temperature are not introduced as a distraction from the serious business of discovering whether intelligence tests measure intelligence. Critics of intelligence testing often condemn out of hand practices and arguments which are the mirror image of practices and arguments which can be found in the history of temperature measurement. If they are acceptable in the one case (and science has no complaints about the way we learned to measure temperature) then why not in the other? For instance, psychologists often complain that there is little proper understanding of intelligence on which its measurement could be based, but exactly the same was true of temperature—it is only in the last century that our ideas of thermodynamics began to cohere into the semblance of a proper theory, and even now there are woeful gaps between the thermodynamic and the kinetic theories of heat and temperature. Indeed, it may be said that it was the invention of the thermometer that made the advances possible which ultimately led to our present understanding (imperfect though it be) of the phenomena of thermodynamics; measurement provides us with a (primitive) method of quantitative assessment, and the results of this assessment provide us with facts which help us to improve our (primitive) theories; in turn, these improved theories enable us to improve our measurements. It is not reasonable to condemn the method of measurement used at any period because it is not perfect; perfection, even if it were possible, is not reached at an early stage in the development of a science. Our methods of measuring intelligence may be primitive, corresponding to the methods of temperature measurement current around the early 1700s; this does not mean that they are not scientific, or that they are not capable of being improved and perfected. The premature imposition of inappropriate criteria of perfection would stultify the development of any scientific discipline.

Let us now return to our quest for a proper definition of intelligence, against which we can measure our tests. To the man in the street, the problem seems a simple one. He commits the grievous error of 'reification'; he thinks of intelligence as something 'out there' which can be compared directly with the results of our IQ testing. But of course there is nothing 'out there'; intelligence is merely a concept regarding which it is nonsense to raise the question of whether it really exists or not, or whether it is appropriately measured by certain tests. A concept is more or less useful; it has no physical existence. We can argue about the degree of usefulness of a concept, or whether the known facts can be better subsumed under a different concept. We cannot usefully argue about whether intelligence exists, just as little as we can argue about whether gravitation exists. Certain phenomena exist which cause us to create concepts like intelligence or gravitation; some people solve intellectual problems, some objects fall to the ground. These facts are based on direct observation or experiment; they furnish us with the basis for conceptualization. But the concepts are our own inventions, and they have to be judged according to criteria different from 'existence' or 'nonexistence'.

How did the concept of intelligence originate? It seems likely that a brief historical recapitulation may help us understand better the nature of this concept. Such a recapitulation may also serve to answer some criticisms that are frequently heard, according to which psychologists secretly and unashamedly stole the notion of 'intelligence' from common parlance, invested it with their own esoteric meaning, and are now trying to pass it off under a false flag, so to speak. Reality is quite different. Plato had already contrasted intellectual-cognitive activities with emotional-conative ones, and Aristotle contrasted the actual or concrete activity with the hypothetical capacity on which it depends, thus introducing the idea of an 'ability'. Cicero translated the original Greek terms into Latin, and coined the word 'intelligentia'. In more recent days, the term 'intelligence' in this sense was revived by Herbert Spencer in his

Principles of Psychology, published in 1870. He regarded it as an ability which enabled its owner to adjust himself more effectively to a complex and ever-changing environment. (The traditional and widespread use of intelligence at the time was equivalent to 'information' or 'knowledge'.) Sir Francis Galton (1822–1911) put the concept on a more empirical level by his studies of famous men and their ancestry, by his advocacy of mental tests, and by his discovery of the twin method for genetic analysis. He was also instrumental in the creation of the correlational method. Following the first or Greek stage, which saw the disentanglement of the cognitive from the non-cognitive side of human nature, this second stage specified more closely the hypothetical nature of the intellectual component, and set about seeking for ways of studying it empirically.

The third stage saw the work of such men as Binet in France and Spearman and Burt in England. Working in the early years of this century, they created proper test devices and they also put forward quantitative theories and suggested ways and means of testing these hypotheses. The concept of intelligence which emerged from all this work was caught well in Burt's famous definition of intelligence as *innate, general, cognitive ability*. Much effort has gone into the detailed testing of these three properties of the hypothetical concept, intelligence; critics often fail to take these studies into account when they condemn the very notion of intelligence. We will discuss the inheritance of intelligence in the next chapter: in this chapter we will discuss the generality of intelligence. These are the two points around which controversy has raged most stormily: the third point has not given rise to much discussion, if only because most workers in this field are quite agreed on the type of problem which could be considered to be 'cognitive' or 'intellectual' in nature. Spearman (1927) tried to lay down in his 'noegenetic laws'* the essence

* Noegenesis is the creation of novel mental content.

of this 'cognitive ability', and these laws are well worth restating, as there is much evidence that they embody valid principles of test construction.

The first is the law of apprehension, stating that a person has more or less power to apprehend outer reality, and inner states of consciousness; in more modern language this might be rephrased to read that a person has the power to encode and transmit information. The second law relates to the eduction of relations: 'When a person has in mind any two or more ideas (using this word to embrace any items of mental content, whether perceived or thought of) he has more or less power to bring to mind any relations that essentially hold between them.' And the third law, relating to the eduction of correlates, states that 'when a person has in mind any idea together with a relation, he has more or less power to bring up into mind the correlative idea'. Spearman's definition of intelligence would be in terms of these three laws; an act is judged intelligent if it involves, and to the extent that it involves, apprehension, the eduction of relations, and the eduction of correlates. IQ problems are almost universally constructed in such a way as to emphasize these abilities.

To exemplify the meaning of these three laws, let us look at a very simple example of a typical IQ test item. Beautiful is to ugly as strong is to ———? Here we must first apprehend the given fundaments, that is the words in question, and their meaning. Next we must educe the relation between beautiful and ugly, namely that of 'opposites', and last we must apply this relation of 'opposition' to the term 'strong' in order to find the correlative term, namely 'weak'. Test items can of course be much more complex than this, but the essential nature of the mental processes involved, as defined by the task set, remains identical.

Consider the following IQ problems which are typical of many others.

1 High is to low as Big is to: small, large, down, fair (Underline correct answer)
2 sheep trumpet cow lion tiger (Underline the odd man out)
3 a d g j m — (Write in the next letter)
4 10 21 43 87 — (Write in the next number)
5 Select the correct figure from the six numbered ones

6 Insert the missing letters

C		I		I		?
E		F		M		?

7 Insert the word that means the same as the words
outside the brackets
EXCAVATION (. . . .) POSSESSION

8 Insert the missing number
16 (93) 15
14 ( ) 12

9 Insert the missing word
SEND (SEED) FEEL
GAME (. . . .) STAY

10   Select the correct figure from the six numbered ones

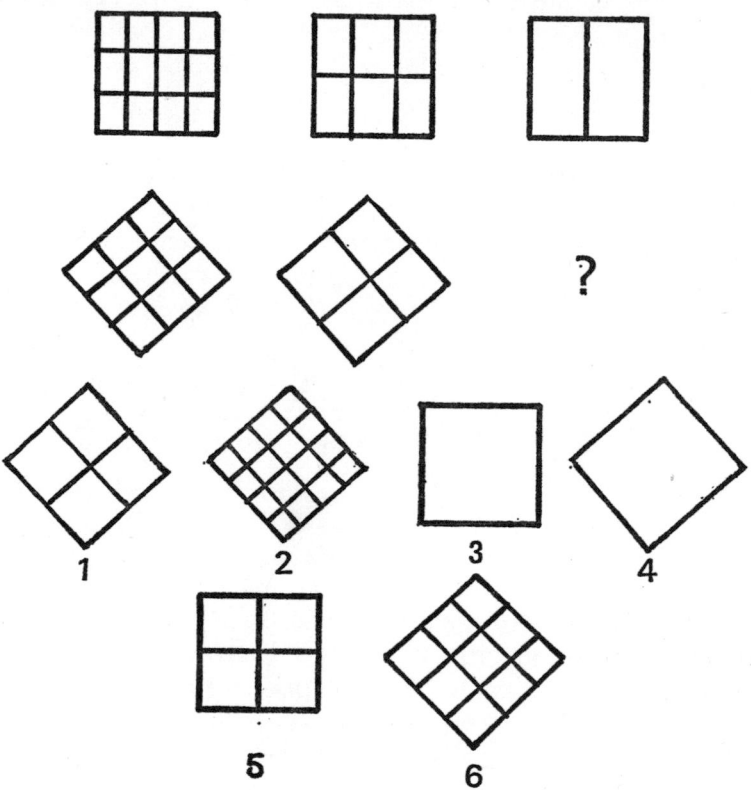

Problems 5 to 10 are taken from H. J. Eysenck, *Check Your Own IQ*, Pelican (1966), where the solutions may be found.

It will be seen that the problems shown all require, in addition to apprehension, the eduction of relations between the elements involved, and the eduction of correlates. The 'fundaments' or elements involved can be anything—numbers, letters, lines, words—as long as these are simple enough not to involve specialized knowledge. Spearman calls this simple rule 'the indifference of the indicator', meaning that the precise nature of the fundaments is irrelevant to the measurement of intelligence. Items can of course be written which look like IQ items, but which are simply measures of specialized knowledge, as for instance the following:

$$Rigoletto : La\ Bohème = Verdi:\ \begin{array}{l} \text{Mozart} \\ \text{Puccini} \\ \text{Wagner} \\ \text{Bizet} \\ \text{Strauss} \end{array}$$

(Put a tick against the correct answer)

Here the eduction of relations and correlates plays quite an unimportant part, and knowledge of which composers wrote each of the works quoted counts for everything. This distinction between 'pure' or 'culture fair' tests and measures of specific knowledge is of course not absolute, and it should not be taken to mean that tests of specialized knowledge may not correlate (sometimes quite highly) with 'pure' tests of what is sometimes called 'fluid ability'. We will return to this distinction again presently.

Spearman (1927), in addition to clarifying the basis of 'intelligent performance' and the cognitive processes involved in it, put forward a theory which has rightly become famous, although in recent years it has been much ridiculed—usually by critics who completely misunderstood his position. He has also been criticized on the grounds that more recent studies have failed to bear out his predictions: we shall have to see if this criticism can be maintained. First of all, however, we must see what Spearman actually said. Put very briefly, he maintained that all tests of intelligence measured the same general mental ability, that is, an ability entering into *all* mental tasks, (usually abbreviated *g*) to varying extent; each test also measured something specific to itself (usually abbreviated *s*). The specific factors, and the abilities underlying them, were quite un-correlated, and they were also quite uncorrelated to *g*. Given this very simple hypothesis, he showed that certain statistical relations would obtain among the correlations between a number of such tests.* Within the inevitable chance errors of measurement, the tables of correlations he and his students produced seemed to come quite close to this ideal, and he concluded that his prediction had been verified.

Critics who believe that Spearman's work and theories have been superseded by later authors usually fail to read the fine print in his book. Among the points to look for are the following:

(1) Among the subjects of the experiment, the whole range of ability in the population should be represented. This is obvious: if we only take a narrow sample of high-grade students, all having very high IQs, then obviously there will be too little variation in IQ to give rise to meaningful correlations. In a similar manner, if we wanted to postulate 'height' as an important variable which distinguished people, we would not try to support this hypothesis by measuring only London policemen all of whom are (or were) over six feet tall.

(2) Each test should be made up of similar items, differing only with respect to difficulty. This again is obvious; if we construct a test containing items each of which represents a different *s*, then we are not dealing with a test in Spearman's sense, and the test would have

* Modern mathematicians would express this set of relations by saying that the matrix so produced would have unit rank, that is show a certain very regular pattern.

to be broken down into parts which were in fact homogeneous.

(3) Different tests must not be too similar, otherwise the *s* factors would overlap. Clearly if one test was constructed on the following format: 'liberty' means much the same as: Whelks Freedom Bookworm Tax Paper, while another was constructed on the following format: Define the following word: 'liberty': —————————, then the two tests, while apparently dissimilar, would in fact share the same *s*—in this case, word knowledge. Admittedly the notion of 'too great similarity' is not a very precise one, but its meaning is unmistakable, and Spearman hoped that empirical research would clarify and make more precise this notion.

(4) Where overlap among the *s* factors was present, Spearman conceded that there might appear 'group factors', such as a verbal factor, or a 'fluency' factor. Admittedly, he was more reluctant to concede the existence of such factors than was Burt, who in the first decade of the century had demonstrated that by having a number of tests depending on verbal material he could isolate a verbal group factor additional to *g*, just as by having a number of tests depending on numerical material he could isolate a numerical group factor, etcetera; but to present Spearman as a simple-minded proponent of an over-simplified theory which took no notice of contrary facts does not give at all an accurate picture of one of the founding fathers of modern theories of intelligence.

Burt and Spearman agreed on the importance of the general factor of intelligence; they disagreed on the importance of group factors such as verbal ability, numerical ability, spatial ability, etcetera, with Burt asserting their great importance and value, while Spearman only admitted their existence reluctantly and tardily. But in 1938, L. L. Thurstone, of Chicago University, published by far the largest empirical study ever carried out: he used fifty-seven tests in all, intercorrelated them, and then factor analysed the matrix of correlations. His conclusion came like a bombshell, for he discovered no trace of *g*, but only a number of 'primary abilities', which was his name for Burt's group factors. It is this study which is still often quoted by critics who believe that Thurstone once and for all laid the ghost of the notion of 'intelligence' as a general, cognitive ability. This, however, is not true—as Thurstone himself was to acknowledge only three years later.

Why did he produce findings so utterly different from Spearman's? There are two main reasons. Spearman and his students had usually worked with school children, representing all shades of ability; Thurstone worked on a specially selected group of university students, showing little spread of intelligence among themselves. He thus contravened the first rule of Spearman's procedure, and this alone would rule out his results as a disproof of *g*. The second reason is statistical; his method of analysis laid down a rule that all factors extracted from the matrix of intercorrelations had to be independent and uncorrelated. (This is the famous rule of 'simple orthogonal structure'.) But this ruled out the discovery of *g* by arbitrary fiat; the data could not disclose a *g* that might be there because the rules of factor extraction forbade it! This seems a somewhat arbitrary way of cutting the Gordian knot, and not one which could gain universal agreement. And indeed, when other psychometrists re-analysed Thurstone's matrix without accepting this arbitrary rule, they all found ample evidence for both a general factor and a number of group factors.

In 1941, L. L. Thurstone and his wife published a similar study in which they had given tests of primary abilities to schoolchildren, a relatively non-selected sample. Immediately Thurstone was brought up against the difficulty which he had previously avoided by using only a very restricted sample of bright students; he found that his method of orthogonal simple structure would not work! Either he had to give up the notion of simple structure (which he was unwilling to do) or else he had to make his factors intercorrelate. He opted for the latter alternative, and concluded, rightly, that Spearman's *g* now emerged from the intercorrelations of his primary factors. Table 2 gives the correlations

Primaries ·	N	W	V	S	M	R
N (Numerical ability)	–	.47	.38	.26	.19	.54
W (Word fluency)	.47	–	.51	.17	.39	.48
V (Verbal ability)	.38	.51	–	.17	.39	.55
S (Spatial ability)	.26	.17	.17	–	.15	.39
M (Rote memory)	.19	.39	.39	.15	–	.39
R (Reasoning)	.54	.48	.55	.39	.39	–
Correlations with *g*:	.60	.69	.68	.34	.47	.84

Table 2   *Correlations between various primary factors and between each factor and* g

among his primary factors: it will be seen that far from being independent, some of these correlate as highly as .55, .54, and .47.

Thurstone also calculated the correlations of each primary ability with *g*; these values are given on the bottom line. As Spearman would have expected, 'reasoning' has the highest correlation with *g*, 'spatial ability' and 'rote memory' the lowest; all this is eminently in line with the theory of *g* and *s*—provided we are willing to say with Spearman that the overlap among the *s* factors gives rise to the various primary abilities. Thus Spearman's most cogent critic has provided us with the firmest support for the existence of a general factor of intellectual ability! Thurstone, who was not only a fine scientist but also singularly open-minded, acknowledged the truth of these observations himself:

This finding [of the correlations between the primary abilities, which resulted in a matrix very close to unit rank] raises the interesting question whether a unique general factor can be determined. Its interpretation here would be that the primary mental abilities are correlated by a general factor which operates through each of the primaries. Each of the primary factors can be regarded as a composite of independent primary factors and a general factor which it shares with other primary factors.

It will be seen that the prevalent notion that Thurstone in some way 'disproved' Spearman (and that consequently the theory of *g* is a chimera) is a figment of the imagination. Thurstone not only did no such thing; except for a very brief period after the publication of his first monograph, he explicitly denied having done so!

The errors which Thurstone acknowledged in his own earlier work have been compounded in the recent studies of J. F. Guilford (1967), of Los Angeles. Guilford has published a theoretical conception of the structure of the intellect *(SI)* in which he postulates three major dimensions along which mental tests can differ. The first dimension consists of *operations* which can be performed on the test material; of these there are five. The second dimension consists of the *products* of such operations; of these there are six. And finally, the third dimension consists of the contents of the test material on which the operations are performed—figural, symbolic, semantic, and behavioural. The $5 \times 6 \times 4 = 120$ cells of the *SI* model are each supposed to give rise to a factor, and in their recent book Guilford and Hoepfner (1971) have claimed the identification of eighty-four ability factors occupying seventy-nine of the cells, with a few cells having two or more factors. There is, as in the case of the early Thurstone, no *g*, and many psychologists not familiar with the intricacies of statistical analysis have taken this to mean that the corpse of Spearman, already slain by Thurstone, has now been drawn and quartered, and finally disposed of with a stake through the heart, buried at an unfrequented cross-roads.

But, alas! Guilford has made the same mistakes as Thurstone had done originally. His groups are nearly always highly selected, that is, subjects are all very bright, or rather dull, so as to restrict the range of talent, and his method of analysis enforces independence among his many factors. He has compounded these crimes by determining the location of his rotations not by some intrinsic relations among the variables, discovered empirically, but by simply forcing them into the closest correspondence with the details of his scheme! No wonder that his scheme has found favour with the statistically naïve, but has been universally regarded with the gravest suspicion among experts in psychometrics and statistics. There is much of interest in his work, of course; his general scheme, regarded as purely descriptive, has given rise to many important new tests. But as a theory of intelligence it fails to account for even the simplest fact, namely the fact that all his tests are positively intercorrelated when allowance is made for the restriction of talent in his populations. The existence of this 'positive manifold', as it is technically called, sounds the death knell for any theory that omits g. Several authors (e.g. Horn) have attempted to reanalyse Guilford's data without forcing the results into a predetermined scheme, and have come up with results which are essentially similar to Thurstone's later work.

In looking back upon Spearman and trying to evaluate his position in the history of science, I would make a comparison with John Dalton, the discoverer of the atom. He originated modern atomic chemistry, although virtually every detail of his conclusions on this subject subsequently proved to be incorrect. Dalton's oversimplifications were very well suited to the needs of chemistry during the nineteenth century, for they were very near the truth. Atoms are not, as he thought, indestructible, but the energies involved are hundreds of thousands of times those of chemical reactions. Atoms of the same element need not have identical weights, as he taught; we now know about isotopes. However, these are so well mixed in nature that large samples nearly always present a constant average weight. Dalton thought that atoms combined in simple whole-number ratios; this too is not true, but the assumption was permissible for the simple substances which were all the early chemists could hope to unravel. Even his values for relative atomic weights and molecular constitutions were for the most part incorrect. Yet it was Dalton, more than any other scientist, who started modern chemistry off on its voyage to the stars. For, as has been well said, 'in devising a general scientific theory, the important thing is not to be right— such a thing in any final and absolute sense is beyond the bounds of mortal ambition. The important thing is to have the right idea.' There can be little doubt that Spearman had the right idea, however much we may quibble about the details of his system. He, more than any other writer, laid the foundations for the study of intelligence as a general mental ability; he was wrong, just as Dalton was wrong, in essentially thinking of his conception as 'indestructible'. Just as the atom has been split, so has *g*; this does not alter by one iota the importance of the concept of the atom, or of *g*.

The first way of splitting *g* we have already noted; it is by way of the various primary abilities which Burt, and then Thurstone and Guilford, discovered. While a case could be made out for regarding them as simply compound *s* factors, produced through the overlap of these *s* factors, this is not the most useful way of regarding them at the moment; a better conception is the hierarchical model originally suggested by Burt, according to which we have the various simple tests at the bottom, giving rise to primary abilities through their observed correlations, while *g* then arises from the observed correlations between the primary abilities. One reason for preferring this view is that genetic studies have shown that not only are individual differences in *g* largely determined by genetic causes, but that in addition genetic causes are active in producing individual differences in several primary mental abilities. These primary abilities are probably much less important socially than are differences in *g*, but it would not do to disregard them completely; we still have much to learn about their nature and relevance to social behaviour.

The second way of 'splitting the atom' of intelligence is in terms of 'fluid' and 'crystallized' intelligence, as Cattell (1971) has so well named the two constituents. The origins of these notions go back at least as far as E. L. Thorndike's distinction in 1927 between altitude and width of intellect. 'Being able to do harder things than someone else can do' is his informal definition of altitude; it clearly involved the notion of scaling problems according to their difficulty. 'Knowing more things than someone else, and being able to do more things than someone else' is his informal definition of width; it clearly involves the prior application of mental ability which is crystallized into knowledge or skill:

> The two things have been somewhat confused in general discussions and in the construction of measuring instruments because, by and large, a person increases the number of things he can do in large part by adding on harder ones, and also because the person who can do the harder ones can on the average learn those which the duller person can learn more quickly than he, and so learn more of them. Consequently what we may call the *level* or *height* or *altitude* of intellect and what we may call its *extent* or *range* or *area at the same level* are correlated and either one is an indicator of the other. It will be best, however, to keep them separate in our thinking.

Thorndike at first thought that altitude would be much more firmly delimited by heredity than width, but he was convinced by his empirical work that this was not so, and later studies have borne him out: 'fluid' ability is no more determined by heredity than is 'crystallized' ability. This is probably not universally true, but applies only in a culture such as ours where literacy is widespread, and where schooling is universal and compulsory; nor may it have been true in our culture in the middle of the last century. But at the moment there is little doubt that these two 'kinds' of intelligence are determined by heredity to much the same extent. This is fortunate because the tests which have been used to determine the heritability of intelligence have combined these two components in different proportions; had their heritabilities been widely different, then different investigators would have reported widely divergent results. We will return to this point in a later chapter.

The notion of 'fluid intelligence' is based on the absence of specially learned ('crystallized') skills and information (beyond such universally available acquired abilities as holding a pencil, making marks on paper, etcetera) and is intimately tied in with the idea of 'culture-free' or 'culture-fair' intelligence tests. The difference between such tests and the more usual type of IQ test is of course not absolute, just as the difference between crystallized and fluid ability is not absolute; these two factors correlate quite highly together, and even the most 'culture-fair' test is not really culture-free, but only relatively so. Nevertheless, the construction of such tests, following the theoretical advances of Spearman's laws of noegenesis, has given us very useful tools for investigating many problems which previously presented great difficulties. Culture-fair tests consist essentially of simple patterns made up of straight lines, such as problem 10 on page 49; no prior knowledge of words or sophisticated concepts is needed to discover the relations involved, or to find the correlates. The hypothesis that such tests are less dependent on cultural factors than are more verbally oriented tests is supported by the fact that such deprived groups as Indians, Eskimos, Puerto Ricans and others do much better on these tests than on the more traditional ones. (Some details of these investigations are given in my book, *Race, Intelligence and Education*.) It should be noted that for research purposes it is not necessary to have pure 'culture-fair' tests; it is sufficient to have tests differing markedly in degree of dependence on crystallized abilities. Results from these enable us to extrapolate beyond the range of our existing tests.

One important point should be noted in this connection. Ideally, a proper test of intelligence should be as culture-fair as possible; yet in actual practice many of the better-known IQ tests are far from culture-fair. Why is that so? The answer, of course, is quite simple. Practically all published tests have been designed for certain practical purposes, such as the selection of children suitable for grammar school education, students suitable for university education, soldiers suitable for officer training, etcetera. In these situations, purity of measuring instrument takes second place to efficiency, and a pure measure of *g* may not be the best predictive instrument for the task in hand. For a schoolboy to succeed in a grammar school, he would be required to have acquired a certain amount of knowledge, to be able to write and read fluently, to have a good vocabulary, and so forth; consequently in designing tests for selection the psychologist may draw on these various skills and items of information in constructing his test items.

The same is true of most selection tests; the most predictive items are usually those which combine fluid and crystallized ability to some degree. Probably these custom-designed tests should not be called 'intelligence' tests at all: it might be better to reserve this title for tests which more closely approach ideals of pure *g* measurement—supplemented, if need be, by other tests measuring knowledge and skills relevant to the selection situation. It is unfortunate that most people only know IQ tests in the guise of selection instruments; many criticisms which can rightly be made of these as pure measures of *g* do not apply to proper *g* tests. A clear distinction should be made between research instruments and applied measures; the latter have multiple purposes, and these should always be borne in mind in formulating criticisms. A test may depart widely from the ideal prescription for a good *g* test, and nevertheless be perfectly adequate as a selection device in a particular situation; it would, however, be preferable if it were not then called an IQ test.

We have noted two different ways of 'splitting the atom'; there is a third, and this may be much more important than the other two. Typically, psychologists in performing their statistical analyses use as their units the scores obtained on complete tests; is this really permissible? The proper unit, surely, must be the individual item; total scores, calculated over a number of items may be very misleading. Consider Table 3, in which the performance of four imaginary people on six equally

| | Problem: | | | | | | |
	1	2	3	4	5	6	Score
John:	R	R	R	—	—	—	3
Mary:	W	R	W	R	W	R	3
Ted:	R	A	R	A	R	A	3
Jane:	R	R	A	R	A	—	3

Table 3 *Hypothetical performance of four people on imaginary test items*

imaginary items has been faithfully recorded. In this table R means an item correctly solved, W means an item incorrectly solved, A means an item attempted but abandoned as too difficult, and — means an item not attempted. It can be seen that each subject of the experiment gets the same total score, namely three. Note, however, that the way this score is made up differs drastically from person to person. John proceeds methodically, getting the easiest three items right, getting none wrong, and abandoning none; he is moderately bright, persistent, and careful in his working. Mary gets some easy items wrong, and some difficult ones right; she is bright, but slapdash and careless. Ted is bright, too, but he gives up too easily; he abandons some easy items which he could have got right if he had tried longer. Jane is less bright, and she too lacks persistence, although perhaps not to the same extent as Ted.

This is a somewhat subjective, interpretative account of what went on; none the less the fact remains that when one looks at individuals performing on IQ tests, such wide variations among people getting identical total scores can often be observed. In what way can it be said that identical total scores, arrived at in such widely divergent ways, are statistically equivalent? We could in fact go one step further and argue that we are wrong in simply scoring a given test item 'correct', giving it one point, or incorrect, abandoned, or not attempted, and giving it no points; we should time the individual and find out how long it took him to solve the item, or how long he worked at it before he abandoned it. Speed is an important aspect of intellectual work; surely it is not reasonable to take no notice of this variable which can so easily be measured?

This line of reasoning has been pursued in our laboratories for several years, and the general outcome of the work done along these lines has been quite clear-cut. It appears that performance on IQ tests can be accounted for in terms of three major mechanisms: mental speed, error-checking, and persistence or continuance. These mechanisms are independent of each other, although this depends to some extent on the instructions given, the motivation present, and other extraneous variables which can be manipulated experimentally. Just as the atom can be analysed into protons and electrons, and many other constituents, so the IQ can be split into these three major components; for serious theoretical work, and I suspect for many practical purposes too, it would seem that we ought to think in terms of speed, error-checking and continuance, rather than simply in terms of IQ. This conclusion may have important social consequences; I shall return to these in a later chapter.

Before leaving this topic of the possible sub-divisions of the IQ, we must look briefly at two such alleged components which in the minds of some critics seem likely to supplant the IQ altogether. These two components are originality or creativity, and the concepts introduced by Piaget. To take the former first, it may be useful to look briefly at the history of this concept. Woodworth in 1918 drew attention to the distinction between what he called 'convergent' and 'divergent' forms of problem solving. The IQ test problems shown earlier are all convergent, that is the relations implicit in each problem converge on one and only one correct solution. But it is possible to think of other problems which do not converge on one correct solution. We might ask the question: How many uses can you think of for a brick? This problem too falls under Spearman's noegenetic laws, employing the eduction of relations and correlates, but there is an abundance of 'right' answers; in fact, it is not always clear what a 'right' answer is, or how many such there might be. Most people would accept the answer: 'to knock a nail in with', but would they accept the answer: 'to sit on?' Clearly the quality of different answers differs grossly, yet the scoring of these tests usually simply consists of adding up the total number of answers given, without attempt to weight them for relevance.

There are of course other difficulties; you can often split an answer up into innumerable sub-answers. 'To kill an animal with' is fine; but suppose you start specifying all the small animals you can kill with a brick—should they count separately? The Spearman school actually experimented with tests of this kind, and showed that there was a factor of 'fluency' involved; this seemed to be made up in part of *g*, and in part of certain extroverted personality traits. Guilford took up this idea some time after the Second World War, and made 'divergent tests' popular. His lead was followed by a number of psychologists (usually educational psychologists) whose enthusiasm for the 'new' type of test was only matched by their lack of scholarly caution.

Foremost among this group were two Americans, J. W. Getzels and P. W. Jackson, whose book on *Creativity and Intelligence* in 1962 caused quite a stir. In it they suggested (and apparently believed they had succeeded in showing) that creativity and intelligence, as measured by IQ tests, were two different and unrelated things. They went on to suggest that Western school systems, by concentrating on IQ and convergent thinking, were neglecting and suppressing the possibly much more important 'creativity' of the pupils. Perhaps the bright and knowledgeable were not, after all, the salt of the earth?

But do the data actually offer any support for this revolutionary doctrine? Note first of all that the five tests of 'creativity' used correlated quite significantly with IQ tests; there is no evidence here of the alleged independence of intelligence and creativity. Correlations averaged about .3 for the boys and girls who took part in the study. But perhaps the 'creativity' tests formed a group factor within general intelligence, a primary ability in Thurstone's sense? Even that seems to be contraindicated by the figures: the 'creativity' tests do not correlate with each other any more than they correlate with traditional intelligence tests! Nor is this outcome peculiar to the Getzels and Jackson study; it seems to be quite universal. Again and again has it been shown by investigators that alleged tests of 'originality' or 'creativity' correlate as highly with IQ as they correlate with each other. Nor could it be said that these tests of 'creativity' have ever been validated, in the sense that adults known to be creative, that is to have successfully created something important in science, medicine, or the arts, score higher on these tests than do quite uncreative people of similar IQ. There is some evidence that these tests do measure certain personality variables which may determine to some degree the way in which a person uses his existing intelligence; the work of Wallach and Kogan, described in their book on *Modes of Thinking in Young Children*, comes to mind. Here too, then, Spearman seems to have been right: when suitable tests of 'fluency' are used, that is tests showing considerable overlap of s factors, these may give rise to a group factor which is appreciably weighted by personality variables, as well as correlating with IQ. It certainly does not rival g in any way, or displace it from its position as the central general factor of intelligence.

While the results of these studies, then, are disappointing to those who attempt to disparage the importance and universality of g, they do have some interesting lessons to teach us. How does it come about that the inconclusive and largely negative results of these badly designed and poorly analysed studies can give rise to a widely accepted myth, a myth which seems quite remarkably resistant to scholarly criticism and factual disproof? Educationalists the world over have accepted the notion of 'creativity' as independent or indeed opposed to orthodox 'intelligence', and arguments have appeared quite widely suggesting that our educational system should be revamped to take more count of this wonderful new discovery. The spirit of Rousseau is clearly still very much alive, and no amount of factual research can kill it; believers will simply disregard the facts which go counter to their predilections, in the well-known spirit expressed by the saying that whenever facts oppose a cherished theory, then 'so much the worse for the facts!' All this does not of course mean that there is no such thing as 'creativity', or that future research may not succeed in measuring it. More ingenious designs and more appropriate experiments may produce different

3

findings. Until that day arrives we may perhaps dismiss as unproven the idea that 'divergent tests' have some special magic.

When we now turn to the work of Piaget (1950), we find a remarkable change of scene; it would be quite impossible to dismiss the long-continued, important and creative work that bears his name on the same grounds. It has to be discussed here because many people have felt that Piaget's theories, and the kinds of tests to which these have given rise, are much more appropriate for the analysis of intelligence than are traditional IQ tests. Thus one enthusiast writes that 'Piaget's approach not only allows an understanding of how intelligence functions, but describes it. Since the interest of Piaget's tests lies in describing the mechanisms of thinking, they permit an individual, personalized appraisal of further potentialities independent of culture.' If this were true, if Piaget's approach enabled us to investigate a child's intellectual abilities independent of culture, and if the resulting measures were also independent of IQ, then indeed we might say that a true alternative to IQ testing had been discovered. But is all this true?

Piaget views development of cognitive functions as going through certain stages—sensorimotor, preoperational, concrete operational, and formal operational; he has derived a large number of ingenious 'tests' or clinical-type procedures for assessing the child's mental development as he moves through these stages, and the various sub-stages into which they can be broken down. These tests are certainly 'culture fair' to about the same extent as Raven's Matrices, or the Cattell tests; Arctic Eskimos excel over white urban Canadian children to about the same (slight) extent as they do on the Matrices, and Canadian Indians do almost as well as Eskimos. Furthermore, formal schooling has no effect on the age of achieving the various component structures and skills that comprise these stages. However, these tests cannot be said to measure something very different from the *g* defined by ordinary IQ tests. Vernon and Tuddenham (in Dockrell, 1970) have shown that correlations between IQ test items and Piaget-type test items are high. In fact, Piaget items have very high *g* loadings, and seem to measure little else but *g*; this speaks equally well for Piaget's insight into child psychology as for the Sparman-type theory of noegenesis which underlies the creation of 'culture-fair', tests of the traditional type with high *g* loading. Along very different paths, these two approaches converge on an identical *g*. Far from being a criticism of ordinary intelligence testing, or providing an alternative to it, the work done with Piaget-type tests strongly confirms the value of the original paradigm. In all ways that have been tested (including heritability) Piaget-type test items behave exactly as one would expect on the hypothesis that they were good measures of *g*; if indeed they 'describe the mechanism of thinking' and are 'independent of culture', then the same must be said of intelligence as measured by IQ tests.

We may end this section by saying quite categorically that the search for alternatives to the general factor of intelligence has failed, just as the attempt to split up intelligence into *independent* parts has failed. 'Creativity' or 'originality' has failed because there is little proof that such a concept is viable, or can be measured with our present-day tests. Piaget's scheme does not provide an alternative, not because of any inherent faults but because of its excellence; it ended up in perfect quantitative agreement with Spearman's conception. It seems we must accept the concept of *g* for the present as closer to the facts than any alternative.

There is an interesting parallel to the position of 'intelligence' in the field of physics, namely the position of the concept of 'hardness'. Few people who have ever slept on bare boards will deny that these are 'harder' in a meaningful sense than is a featherbed; nevertheless the concept of 'hardness' presents difficulties to physicists which are similar to those presented by 'intelligence' to psychologists. The concept of hardness itself is obviously derived from our experience with the fact that a given material can be drilled, sawn, indented or abraded with greater or lesser ease than another material. It is further intuitively obvious that the harder of two substances will scratch the other, will resist wear better, or suffer less damage when struck by a third material harder than either. Yet these concepts do not in fact involve a single physical property, but various combinations of several properties. As a result there is more than one kind of hardness. Abrasion hardness, for instance, depends to a large extent on the properties of the surface so that the way the material is prepared and the effects of corrosion by the atmosphere can both be important. Scratch hardness involves a combination of plastic flow and fracture characteristics; the shape of the scratching tool can play an important part. Plasticity, or lack of it, largely determines indentation hardness, though brittleness may also make for easy indentation. Thus lead and talc are both soft materials, but for different reasons. Lead flows readily when indented; talc offers little resistance to fracture and crumbles easily. The range of properties loosely considered as 'hardness' is so wide that it is impossible to arrive at a concise definition that includes every characteristic. Any precise definition of hardness depends on a particular measure of assessment. Thus hardness is what tests of hardness measure—we seem to have returned to our starting point! Yet note that no physicist would say that because of these difficulties of definition it is meaningless to speak about hardness, or unscientific to investigate different materials from the point of view of their hardness. The Mohs scale (what scratches what) was widely used since it was introduced in 1822 by the geologist F. Mohs, and is in fact still used in conjunction with other physical characteristics in the preliminary identification of minerals. IQ tests have much in common, from the point of metrology (the science of measurement) with Mohs' scale; this is meant as neither praise nor criticism, but simply as a statement of fact.

Critics have for many years complained that the use of the term 'intelligence' for the mathematically derived concept of *g* is fallacious and adds a halo of irrelevant and probably erroneous meaning to an otherwise sterile conception. Some psychometrists have felt like acceding to this request, and prefer to talk about factors in terms of simple classificatory schemes, using either the letters of the alphabet (*g* for general, *n* for numerical, *f* for fluency, *v* for verbal, etcetera) or a series of numbers. To do this simply avoids dealing with a very real and meaningful problem; the question still remains of whether the *g* the psychometrist talks about resembles in some important way the intelligence the man in the street deals with. Newton met the same problem when in his *Principia Mathematica* he dealt with Mass; he points out there that what the physicist means by this concept, and what the man in the street means by it, are two very different things. He adds that it is a 'vulgar error' to confuse the two, but he does not advocate referring to his 'mass' by a simple letter, *m*. The term intelligence originated, as we have seen, in departments of psychology and philosophy; it is by no means clear why we should give up using it because it is suggested that the man in the street uses it in rather a different sense. But is it in fact true that the man in the street does use the word in a different sense to that given it here?

This is a difficult question; no one has ever asked a random sample of the population what 'intelligence' means to them, or how they would define it. In a small sample I once questioned such definitions as the following were frequently given: bright, clever, able to think clearly, not thick, all there, gifted, do well at school, capable, plenty of ability, successful with ideas, knowledgeable, full of ideas. It is not obvious that these views conflict in any way with the definition given by Spearman or Burt; there seems to be much overlap. Certainly most people associate intelligence with success in school, or in learning generally, and as already pointed out, there is no question that IQ tests predict success at school, in college, or at university very well. Furthermore, when teachers are asked to rate their pupils for intelligence, these ratings always correlate quite well with IQ scores; there are occasional discrepancies, but on the whole correlations are positive and quite high. So far we see no reason to discard the term from our scientific vocabulary.

We may go further than this. It is possible to rank occupations in order of social approval, in the sense that at the top we can put professions which people regard highly, value correspondingly, and would like their children to enter; these are usually considered to require 'brains'. Similarly, we can put at the bottom occupations which people tend to regard as less valuable, value less highly, and prefer their children not to enter; these are usually considered to require more 'brawn' than 'brain'. Many different investigators have drawn up lists of occupations, and have asked random samples of the population to rank these in order of prestige, or social value, or importance; these rankings, in many different countries, have been found remarkably similar. We can now test people in these various occupations and discover whether in fact those with occupations near the top of the prestige ladder do in fact have higher IQs than those lower down. If *g* does in fact have much to do with 'intelligence' as popularly understood, then there should be a fair deal of correspondence.

Table 4 shows results from a number of empirical studies carried out in different countries; there is usually considerable agreement between investigators on the results, so that no more detailed presentation is required. There is a rough 50-point range from the highest group—higher professional, top civil servants, professors and research scientists, having IQs averaging 140—to the lowest—labourers, farmhands, miners, unskilled

140	Higher professional; top civil servants; professors and research scientists.
130	Lower professional; physicians and surgeons; lawyers; engineers (civil and mechanical).
120	School-teachers; pharmacists; accountants; nurses; stenographers; managers.
110	Foremen; clerks; telephone operators; salesmen; policemen; electricians; precision fitters.
100+	Machine operators; shopkeepers; butchers; welders; sheet metal workers.
100−	Warehousemen; carpenters; cooks and bakers; small farmers; truck and van drivers.
90	Labourers; gardeners; upholsterers; farmhands; miners; factory packers and sorters.

Table 4 *Mean IQ of different professional and occupational groups*

workers, having IQs averaging 90. (It would be possible to include another group below the last-named, made up of unemployables, drifters, inmates of institutions, etcetera, with an IQ around 60 to 70, but this seemed a task of supererogation.) The main point is surely clear; there is an almost perfect agreement between the social prestige of an occupation and the mean IQ of those in that occupation. This is not perhaps surprising; those at the top of the scale have professions which require high intelligence and lengthy schooling and training, and as we have already seen, schooling depends very much on high IQ. Nor has this association between brains, education, and occupations of high social status escaped the notice of the man in the street; the wide desire for better education, and the high hopes associated with successful schooling, are a well-known feature of the contemporary scene.

It is easy to misunderstand the scale of social prestige given in the table, or to look askance at any form of stratification which puts an élite at the top of the pecking order. Let it be said at once that this is not a scale of social usefulness; society could not exist without miners, or nurses, farmhands or secretaries, while it could exist perfectly well without research scientists (at least for a while)! Nor is it a scale of monetary and financial success; typically in the USA a university professor might get less pay than a master plumber, and even in the UK a secretary, nurse or teacher would get less than a miner, butcher or carpenter.

In his book, *Passion to Know*, Wilson makes the interesting observation that when the atom bomb was being constructed at Los Alamos during the war, with the most able and brilliant band of physicists ever to be assembled working day and night to solve the myriad problems which arose in the course of pushing back the borders of the unknown, 'the scientists, on their university pay, received half the pay of the union technicians who worked for them'.

The correlation between IQ and pay is less close than that between IQ and social prestige; this situation is probably different in the USSR where research scientists and professors not only have high prestige, but also salaries more in line with their IQs (at least among the really successful ones). However that might be, there is no doubt that success in life, defined either in terms of income or of social prestige (and thus defined essentially in terms of what the man in the street thinks) correlates quite well with IQ (although not perfectly, as we have seen); hence IQ does clearly measure much the same sort of thing as the man in the street means by intelligence.

These are correlations *between* occupations. One would expect there to be a considerable spread of IQ *within* each occupation, and hence a somewhat lower correlation between individuals' IQ and occupational status. There is no doubt about the marked spread of IQ within occupations; this is much more marked among the lower than among the higher occupations—not unexpectedly, as it is easier for a bright person to enter a nondemanding occupation (through, for example, indolence, sickness, or bad luck) than for a dull person to surmount the many hurdles which society has interposed in the form of examinations, working oneself up in a job, etcetera. Roughly speaking, correlations between .6 and .7 can be found between schoolboys' IQ and their occupational status later on in life.

Even when dealing with samples whose members had had the same schooling and were in the same line of work, IQ still correlated with earnings, although of course not so highly. IQ is correlated more with the type of job finally achieved, than with the earnings within this particular type of job. This is due in part to the obvious restriction of range of IQ; people in a given job range around the common mean, and do not usually depart far from it. There is also the influence of trades unions and other negotiating bodies; these prefer uniform salaries for everyone doing the 'same' work, irrespective of how well or how poorly they do it.

All these figures fit in very well with the hypothesis that IQ is important in determining a person's standing in our society, and that the main influence of IQ on status is indirect, that is through education. As we will see later, educational status is itself very strongly determined by IQ, and to a large extent independently of social class.

All this can be put in numerical terms. First we have the Barr scale of occupations: this was drawn up by a number of psychologists who rated 120 representative occupations with respect to the grade of intelligence required in each one for ordinary success. Second, there are the results of a large-scale public opinion poll, undertaken by the National Opinion Research Centre (NORC), in which the prestige ratings of a great number of occupations were established. Last, we have ratings of socio-economic status (SES) as assigned officially in the Census of Population of 1960 to each of hundreds of listed occupations on the basis of average income and educational level prevailing in the occupations. Prestige (NORC) ratings and intellectual requirements (Barr scale) correlate .91; prestige and income (Census data) correlate .90; intellectual requirements and income correlate .81. There is thus a close relation between the intelligence needed in an occupation, its social prestige, and the income and education of the people in it. If we regard income and prestige as having some importance, then it is clear that intelligence precedes occupational choice, and is thus causally implicated in the other two variables.

Perhaps the most impressive evidence that the gifted child (in terms of IQ) grows up to become the successful adult (in terms of financial and prestige criteria, such as the ones discussed above) is given in Terman's famous study of gifted children in the United States (defined empirically as children with IQs of 140 or above). Terman's group consisted of 857 male and 671 female children, and they have been followed in their careers for over 35 years; a great deal is known about their success or failure. The mean IQ for the two sexes was in fact 151 at the original testing, and a great deal of further information was obtained about each child, his family background, home environment, etcetera. The follow-up showed 'that the superior child, with few exceptions, becomes the able adult, superior in nearly every aspect to the generality'. Repeated testing during these years showed that the gifted children tended to maintain their superior IQ. Their educational record was a distinguished one; more than 85% entered college and almost 70% graduated. Their occupational record too is very superior; clearly we must here distinguish between the men and women, as the status of the latter tends to be defined to a considerable extent by their choice of marriage partner. The women have certainly produced considerable evidence of achievement; seven are listed in *American Men of Science*, two in the *Directory of American Scholars*, and two in *Who's Who in America* (all before reaching the age of 43!). They have published 5 novels, 5 volumes of poetry, and some 70 poems variously published; 32 technical, professional, or scholarly books; about 50 short stories; 4 plays; more than 150 essays, critiques, and articles; and more than 200 scientific papers. At least 5 patents have been taken out by gifted women. As for the men, 86% are found in the two highest occupational categories (professions and higher business). 11% are in smaller retail business, clerical, and skilled occupations. Farming and related occupations account for nearly 2%, and the remaining 1% are in semi-skilled work. The list of the achievements of these men in science, business, and the arts would fill many pages. There can be no doubt whatever that on any current measure of social achievement, the gifted children came up trumps. There were a few failures, but these could be explained largely in terms of failures in mental health, emotional stability and social adjustment—a timely reminder that IQ, while extremely important in determining success, is not the only factor that is important; personality variables, too, come into the picture.

It would be possible to go on for many pages quoting evidence of the relation between IQ and success in many different spheres of life, but there would be little point in painting the lily; the evidence is so overwhelmingly strong that no one can possibly doubt the strength of the connection. The Terman study has been mentioned, not so much because it is the most interesting and scholarly of its kind, but rather because it gives us a chance of testing a hypothesis which is likely to have occurred to many readers—is it possible that these gifted children had high IQs, and were successful, because of the social environment provided by their parents, who were also presumably intelligent and socially successful? The parents of these children were certainly brighter and more successful than the average (although much less so than their children); what would we expect the children of our gifted group to be like, seeing that their parents were even more favoured? This will be discussed in a later chapter.

Many people feel that the psychometric model of intelligence lacks bodily substance; it seems intuitively easier to measure the length of a table, or the temperature of a room, than to measure something insubstantial like intelligence. Yet there are more similarities than dissimilarities between the measurement of intelligence and the measurement of temperature; in either case, we are concerned with behaviour—in the one case with the behaviour of human subjects in a problem-solving situation, in the other with the behaviour of molecules jostling each other at varying speeds. Nor would it be true to say, as some critics have stated, that the cases are different because in the case of intelligence we have no good theory on which the method of measurement is based. As we have seen, there was no good theory for the measurement of temperature either when traditional methods were being developed. However, it would be true to say that if intelligence is truly to be regarded as a biological entity, subject to the well-known laws of heredity, then we would expect it to have 'a local habitation and a name'—in other words, we would expect to be able to find certain properties of the central nervous system or the cortex which could be shown to be correlated with intelligence, and which might theoretically be considered to underlie intelligent activity. Recent years have brought us much closer to the point where we could claim to have made at least a beginning in the search for such biological underpinning.

From the very beginning of the intelligence testing movement, efforts have been made to relate IQ as tested to some brain features which might be thought of as causally involved in the production of intelligent behaviour. Brain size itself is of course the first variable that suggests itself, and indeed there does appear to be a slight but definite relationship. A dozen or so studies have been made of the relationship between intelligence and cranial capacity. In all instances, the correlations have been positive, although small, ranging from .08 to .34. Of course, head size is a very rough guide to brain size (because of different skull thickness, differing proportions of white and grey matter, differing body size, different arrangement of convolutions, etcetera) and even brain size does not take into account the number of cells per cubic inch and other microscopic and macroscopic details of the cortex. One might say that no really serious effort has in fact been made to relate IQ and brain anatomy, so that the positive but slight correlations found so far are encouraging but not sufficient to give us more than the most indirect of hints as to the real extent of any relations there might be. Perhaps it is the need for interdisciplinary and longitudinal research that has put off researchers; whatever the cause, very little is known about this aspect of the body-mind relation.

Another physiological variable which was thought to be related to intelligence was speed of neural transmission, and early investigators attempted to measure it through the latency* of certain reflexes, such as the patellar tendon reflex—without much success. Reaction time measurement too was found to be unproductive: the speed with which a person could react to a signal did not correlate at all highly with IQ. However, if you complicated the experiment by having several signals and several different keys to press in response, the increase in reaction time over simple reaction to a single stimulus was found to be related to IQ—the brighter the person, the less increase in reaction time did the increased load of information processing produce. It would seem as if reaction time itself had no relation to IQ, but that once the coding and processing of information came into the picture, people with higher IQ could deal with

* Latency can be defined as the time elapsing between the giving of a stimulus and the sensory response.

C

this more speedily than people with lower IQ. This
makes good theoretical sense, as well as being in accord
with common sense. It also agrees well with work on
the 'evoked potential' on the electroencephalograph, to
which we will turn after considering one other recent
piece of research.

During the Second World War, a junior sergeant
tester in the British Army, engaged on statistical analysis
of test data, found himself with nothing to do. Following
the Army principle that one must at least appear to be
busy in case somebody notices one's idleness and gives
one something even more unpleasant to do, he started
correlating any sets of data he could lay his hands on.
He was rather surprised when he discovered a significant
correlation among these meaningless data, and when he
went to see just what it was that he had correlated, he
found that it was IQ, on the one hand, and number of
teeth missing on the other. This correlation of —.63 was
quickly verified by correlating the same variables on
other samples, and still constitutes the highest correlation between intelligence and a physical feature of the
human organism that has ever been discovered. Unfortunately the causal chain is unlikely to go from the
possession of teeth to the possession of intelligence; social
class (at that time at least, that is before the National
Health Service had come into being) determined very
largely the dental care lavished on a person's teeth, and
as we have seen social class is also highly correlated with
intelligence. There may also be a possibility that more
intelligent people, irrespective of class, take greater care
of their teeth.

Of more interest to many investigators have been the
electrical 'brain waves' recorded on the electroencephalograph, and in particular the alpha wave.

The sad truth is, however, that in adults there is no
simple relation between alpha rhythm and ability
measures. With children there are correlations ranging
from .3 to .6, but these are with mental age rather than

with IQ, that is they seem to be related to maturation. This whole field is ripe for more intensive study.

One aspect of EEG work has been rather more promising, and has been more fully developed in recent years, namely the so-called evoked potentials, that is the waves of negative and positive electricity evoked by a sudden stimulus. (See Figure 2 for a typical example.) These waves are quite characteristic for a given

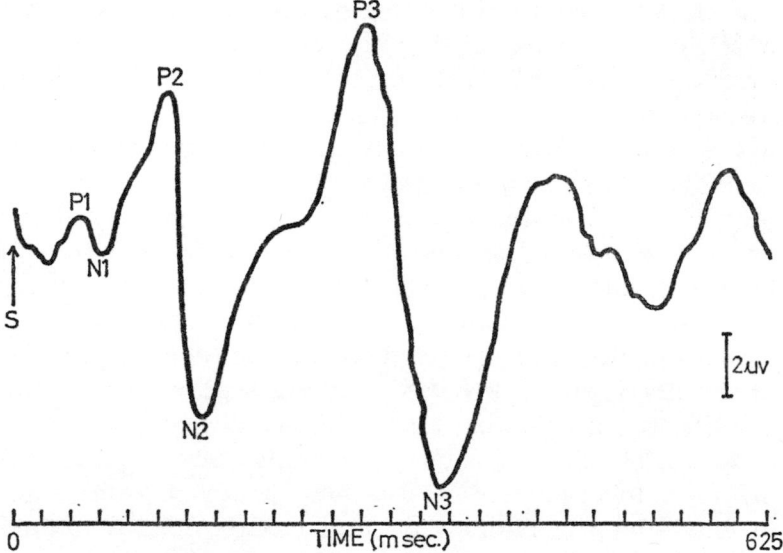

Figure 2 *Illustration of evoked potential on the EEG*

person, both in shape and latency, and it has been suggested that they are connected with referring input to analysing mechanisms, and with the establishment of memory engrams; possibly they might be measures of the 'speed of processing' of bits of information as these enter the cortex. Cattell (1971) points out that:

> it seems reasonable to suppose that they are concerned not only with memorizing but also with the *evaluation* of the stimulus, its referral to the sorting in the sensory area, and also with the eduction of

relations. For they appear when relations are demanded with other sensory areas, as when one presents a standard perceptual intelligence problem. Now a smaller total cortical apparatus, like a smaller computer, might be expected to take longer to process a fixed number of relations up to the required level for solution, as presented by a standard test problem.

Ertl, a Canadian psychologist, reported observations which did in fact show a correlation of about −.7 between intelligence and a latency measure (taken from stimulus presentation to the third wave crest); in other words, quicker waves characterized bright subjects. In spite of the high reliability of evoked potential latencies such a high relationship is intrinsically unlikely, and hence the original observation met with incredulity. However, later work showed that correlations of rather more modest magnitude (that is in the neighbourhood of −.3) could be reproduced with some reliability. Certainly the results leave little doubt that we can now identify at least one of the physiological features correlated with intelligence, and the high reliability of this measure, together with the difficulty Ertl has reported in changing the latency by any environmental manipulation, suggests that the pattern is very likely to have a strong hereditary component. Unpublished work by J. Rust has verified this surmise.

Work in our own laboratories has given strong support to the results reported by Ertl with some interesting additional findings. Using 93 adults, randomly sampled, Elaine Hendrickson administered a test of intelligence; she also determined latencies and amplitudes of evoked potentials in response to sounds.

Her results, in brief, showed that intelligence was correlated both with the latency (negatively) and the amplitude (positively) of the evoked potential; both sets of correlations averaged around .4, and latency and amplitude were themselves uncorrelated. We may thus add their contributions and say that IQ correlates about .6 with a combined latency-amplitude score on the EEG. The evoked potential scores were found to be very reliable and reproducible. Verbal IQ correlated better with evoked potential than did spatial IQ; this is interesting as verbal IQ tests tend to have higher *g* loadings than spatial IQ tests. These correlations are somewhat higher than those previously reported; this is probably due to the fact that auditory stimuli were used rather than visual ones, and there are good reasons to suspect that visual stimuli are less reliable, and produce more arte-facts, in EEG work.

What precisely does the EEG-evoked potential measure? It has been suggested that differences in latency particularly are a measure of 'neural efficiency'; it has also been considered that such latencies measure the speed of information processing, or that of encoding information. There is probably some truth in all these suggestions; it cannot be said at the moment precisely which of them is more correct. Work is of course proceeding in this area, and we should soon have a better understanding of the physiological processes underlying IQ. Even as things stand, however, these studies are important in two ways. In the first instance, they demonstrate without any question that intelligence does have a physical basis which has been found to have strong heritability. In the second place, they suggest strongly that *g* is more than a statistical conception; tests of many different 'primary abilities' have been found to correlate with the latency of evoked potentials, which tends to contradict the idea that there might be some 120 or so independent intellectual factors, as Guilford suggests. If these factors were truly independent, then only one would be expected to correlate with evoked potential latency; in practice, all the factors tested have been shown to have such correlations, and these tend to be proportional (in a rough-and-ready way) with the g loadings of the tests. Here then we have another support for the theory of a general, cognitive ability underlying intelligent behaviour.

In this chapter we have examined some of the evidence relating to the problem of whether IQ tests really measure intelligence, as the term is usually understood. We have drawn attention to the erroneous assumptions which are made by many people in connection with this question; 'intelligence' is not some *thing* out there with which we can compare our measures and decide whether or not the two correspond. Such reification of scientific concepts is inadmissible; all we can do is to construct a theory or model of 'intelligence', preferably a quantitative one, and make testable deductions from this model. If these deductions stand up to experimental testing, then we may say that our theory or model seems to be along the right lines. The Spearman-Burt-Thurstone model is in fact a quantitative one; this accounts for some of its difficulties. Most sociologists, psychiatrists, educationalists and psychologists are largely innumerate, to use Lord Snow's term; they tend to think in verbal terms only, and fail to criticize the model on the only grounds which are really relevant to its correctness, that is quantitative ones. Some critics even take pride in their ignorance, rather like the French physicists who criticized Newton's work without bothering to learn his calculus. Such verbal criticisms are largely meaningless; the only ways to disprove a quantitative model are to make quantitative deductions from it, test them, and show them to be erroneous, or to demonstrate that different parts of the model are incompatible, that is call for different quantitative values. Neither of these two methods has so far produced any serious difficulties for the model. It is also interesting to note that most of the vociferous criticisms published are found, not in serious scientific publications but in popular papers, journals and books.

By the same token, it is difficult to write a popular

account of the model, which will be intelligible to non-experts, but which will also give some impression of the underlying quantitative system of proof and deduction. I have tried here and there to suggest in words the kind of quantitative thinking which underlies the theories, deductions and arguments of the people who constructed the model under discussion, but words are no proper substitute for numbers; as the great Lord Kelvin said: 'One's knowledge of science begins when he can measure what he is speaking about, and express it in numbers.' Or, as Clerk Maxwell put it, 'We owe all the great advances in knowledge to those who endeavour to find out how much there is of anything.' This is the true achievement of the early workers in this field; they showed us a way of finding out 'how much there is' of intelligence. Our measurements are far from perfect, and many improvements can and undoubtedly will be made in them; nevertheless, the groundwork has been well and truly laid. For the innumerate there is no way round the unappetizing choice of either having to accept conclusions at second hand, or else becoming properly efficient at statistics themselves. To indulge in purely semantic and usually irrelevant criticism of a model only dimly understood, and in ignorance not only of the vast number of facts which support it but even of the methods used to construct it, seems somewhat irresponsible; it calls to mind the Aristotelians who refused to look through Galileo's telescope for fear of what they might see.

What has been said above does not of course apply to constructive criticism; there is no implication that the model as at present constituted is perfect and above any criticism. There is no doubt that Thurstone considerably increased the precision with which g can be located in multi-factor space; there is no doubt that Burt improved the simple Spearman model by insisting on the importance of group factors and a hierarchical arrangement; there is no doubt that Guilford contributed valuable information on possible further factors at the primary abilities level. But these are all constructive contributions which improve the model; they do not eliminate it or substitute a different one. There are many suggestions presently being tried out which may hopefully improve our methods of measurement, but these again are meant to improve, not destroy the model.

As one example, consider 'tailored testing'. This approach is based on the fact that the typical IQ test is very wasteful of information. A test containing, say, 100 items differing very much in difficulty will contain some 90 items which are irrelevant for any one subject; many items will be too easy to be discriminative, and many others will be too difficult. Real measurement will only be carried out by a few items in that region of difficulty where the subject's chances of a correct solution are around 50%. Now let us enter a large number of items of known difficulty level into the memory banks of a computer, which is programmed to present these to the subject, who reacts by pressing a numbered key: the computer determines whether the answer is right or wrong. The first problem is half-way up the difficulty ladder; if it is passed, the next item to be presented is at the 75% level of difficulty, while if it is failed, the next item to be presented is at the 25% level of difficulty. Depending on whether the second item is passed or failed, the third item will be more or less difficult, and the computer will continue to select items in such a way as to 'zoom in' on the crucial set of items which will really test the subject's IQ, avoiding all the useless items which are below or above his true level. In this way, by individually tailoring the test to each subject, we could obtain more reliable information in much less time than we do at present.

Our model, then, is far from perfect and is subject to constructive criticism; similarly, the methods of testing used can be considerably improved by 'tailored testing' and other methods. Furthermore, there are undoubtedly anomalies present in the model; no scientific theory or model has ever been without anomalies. But the model is still the only one in existence to explain the known facts. Until a better model is put forward which not only explains the known facts as well as the existing one, but also explains other facts not at present included within the model, or predicts facts which the present model cannot predict, we have no choice but to retain our present model and try to improve it in accordance with our increasing knowledge. That is the usual way of science, and it is difficult to see why we should depart from it in the case of intelligence.

CHAPTER 3

Intelligence and heredity

What was said at the end of the last chapter applies with equal cogency to the subject matter of the present one. We have a model of inheritance of intelligence (and of personality) which is far from complete, perfect and all-embracing, but which accounts for a large number of apparently disparate facts: constructive criticism of this model is more than welcome, and indeed is helping to improve it all the time. This model is quantitative, and in due course I shall attempt to illustrate just what is meant by this, and in what way this property immeasurably strengthens the model and enables it to be used with much more confidence than could ever be the case with a purely semantic model. The frequently heard criticisms of the model—or rather, of some conclusions to which it leads—pay little or no attention to these quantitative properties: one might say that the critics do not in fact address themselves to the model at all, but talk at right angles to it.

The very nature of the problem is frequently misunderstood. We have already quoted Thoday's remark in Chapter 1 that 'every character is both genetic and environmental in origin . . . Genotype determines the potentialities of an organism. Environment determines which or how much of the potentialities shall be realized during development.' Our concern is not with the existence of such characters, but with the causes of variety in a population. We are asking how we can account for the diversity in the expression of the character which can be observed in the population. There obviously are considerable individual differences in all psychological variables that we can measure; the question is to what extent these differences are caused by genetic and environmental factors.

The term 'variance' is often used to give greater precision to this notion of variability, individual differences, etcetera. It has a precise technical meaning. Consider Figure 3, which is a schematic drawing of the distribu-

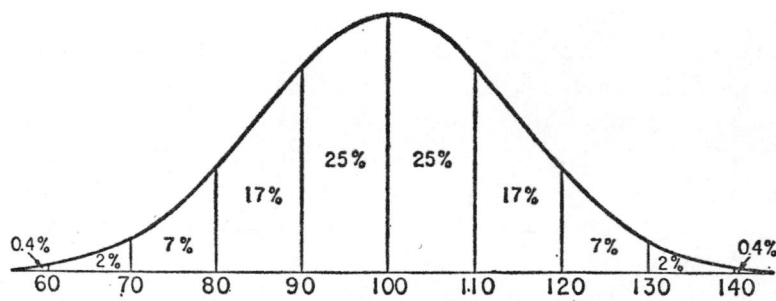

Figure 3   *Normal curve of distribution of intelligence*

tion of IQ in the population. (The true values differ slightly from those in the drawing, particularly at the lower end, where there is a bulge below IQ 50, produced by certain types of mental defect.) The idealized curve is sometimes known as the Gaussian curve of distribution, and it shows roughly the number of people of a given IQ to be found in the population—thus there are about 50% between 90 and 110, 2% between 130 and 140, and so on.

The bell-shaped curve changes direction half-way up and half-way down, from being concave to being convex in its curvature; if we locate the exact spot where this change takes place we can identify one of the two variables which serve to give a mathematical description of the curve. In the case of IQ this point lies at 115 IQ (or 85 IQ, on the other side) and the distance of this point from the mean (15 points of IQ) is known as the standard deviation (SD). Variance is the square of this value (225 in our case) and it serves as a measure of the total variability in IQ which we have to account for.

This total variance can be split into variance due to heredity and variance due to environment; the former is often referred to as 'heritability', and is symbolized as h^2. Heritability obviously cannot be determined for any individual since h^2 is conceived of as a proportion of total variance, and variance is derived from a group of individuals—usually a sample of a population, rather than the whole population. Being a characteristic of a population, h^2 is unlikely to be constant; it may vary from time to time, and from place to place. Making education more uniformly accessible, for instance, will make the effect of heritability on g proportionately greater as the proportion due to environment decreases; restricting education to the favoured few will have the opposite effect.

Thus h^2 can be determined for a given population at a given time. We have a good idea of what it is in England, or in the USA, at the present time; we do not know what it was in these countries a hundred years ago, or what it is in China or India at the present time. We could of course find out the answer to the second part of this question by experiment, and we could get a reasonable approximation to the first part of the question by extrapolation from known or estimated values. But this is hardly the point; what has to be stressed is that heritability is a group characteristic, and that it depends on the particular group which is being studied. In what follows we are concerned with heritability in the Western world, and figures quoted below should not be taken to refer to countries behind the Iron Curtain or to underdeveloped countries.

There are many elements which enter into any equation written to determine h^2, and it is important to realize that until quite recently several of the terms in this equation could neither be measured nor estimated. Thus when some geneticists and psychologists interested in this problem declared that the available evidence showed quite clearly that heredity was important in determining individual differences in IQ, but did not enable us to deduce any quantitative values for the degree of heritability, they were not being obscurantist but simply expressing their unwillingness to believe that because they could not measure parts of the equation, those parts must be unimportant and small in value.

Such terms as the interaction between genetic and environmental factors were not included in the original formulae; as it happens, they turn out not to be very important, and to that extent those geneticists and psychologists who were willing to accept the original estimates based on the older formulae were justified. But fortunately we do not have to make decisions one way or the other any longer; we now have mathematical formulae which enable us to measure the various components of the total variance, and we can construct a proper model of hereditary determination of IQ, a feat which even at the time that Jensen (1969) wrote his famous paper was beyond our capacity. Critics sometimes quote with pleasure well-known geneticists or psychologists who state more or less explicitly that it is impossible or meaningless to give a precise quantitative estimate of heritability; they seldom realize that these quotations were fully justified at the time they were written, but do not reflect present knowledge or present conditions. It is important to understand the changes that have taken place in this field in the last few years, largely due to the work of Jinks and Fulker (1970) at the University of Birmingham. It would be true to say that their work has revolutionized the field.

The only formulae for deriving heritability indices which are familiar to most psychologists were originally derived by K. Holzinger in 1937, for a very well-known book entitled *Twins—a study of heredity and environment,* co-authored by H. Newman and F. Freeman (1937). These formulae have been very widely used but they do not constitute a proper model of inheritance, nor do they take into account many important variables. Here is what Arthur Jensen (1973) had to say about Holzinger's formulae: 'Thanks to Holzinger's simple and often cited, but theoretically nonsensical, heritability index, all the heritabilities reported in Newman *et al.* purely and simply are wrong, that is they do not represent estimates of what any geneticist has ever meant by "heritability".' Thus the criticisms made of studies using these formulae are quite justified, but this should not lead us to believe that more recent formulae and models are subject to the same weaknesses. In particular, the careful analyses performed by Jinks and Fulker (1970) and their colleagues at Birmingham, are models of what genetic work should be. Criticisms which apply to the older studies are irrelevant to these more recent ones, or to the re-analyses carried out on some of the older work; it was the analysis which was faulty in these studies, not the method of collecting data. It is these quite recent methods and models which enable us to check directly on such variables as assortative mating,* dominance,† environment-heredity interaction, etcetera, and which justify us in asserting that the heritability of intelligence is approximately .80. Critics of this figure would have to tangle with the geneticists who have wrought this revolution in analysis, not with psychologists who may claim to understand, but not to have originated, this important advance.

We must now look at the many different ways in

* Assortative mating is where choice of sexual partner is not random. Positive assortative mating is where couples tend to be more alike than chance might permit; negative assortative mating takes place when couples are less alike than chance factors might allow. Assortative mating for intelligence is strongly positive, spouses correlating about $+.5$ with each other.

† Humans inherit 23 pairs of chromosomes, which are the carriers of genes. The two genes at identical loci on a pair of chromosomes are called alleles; often their effects are simply additive. Sometimes one allele is *dominant* over the other, which is recessive; the dominant form finds expression in the physical or mental trait shown by the offspring. Such dominance may be complete or partial.

C

Intelligence and heredity 91

which it is possible to test the hypothesis that heredity, as well as environment, plays an important part in producing differences in IQ. It will be seen that these methods are quite variegated, using distinctly different approaches; this is one of the strengths of the theory, of course, because it enables us to look at the phenomenon from many different points of view.

The first point to note in this connection is that the methods used fall clearly into two quite different groups. If the observed variation in IQ is made up of two variables, due respectively to hereditary and environmental factors, then we can measure heritability in two ways. We can carry out genetic studies (for example, on twins) in which we try to assess the genetic variance directly, or we can try to assess the contribution of environmental factors and arrive at maximum possible heritability by subtracting the environmental contribution from unity (because heredity and environment between them cannot account for more than 100% of the total variance). This point is somewhat oversimplified because we have neglected test unreliability (which really fits into neither the hereditary nor the environmental variance portion) and interaction between heredity and environment; but the former is easily measured, and the latter has in fact turned out to be relatively unimportant. For the purpose of this discussion we may conveniently disregard these complications. (If we were concerned with accurate estimates of heritability, we would of course have to pay attention to complications of this kind.)

We shall consider first of all direct attempts to estimate heritability, and then turn to the measurement of environmental influence through the study of foster children, orphanages, etcetera.

We turn first to the studies of identical twins brought up in separation. As is well known, there are two kinds of twins—identical or monozygotic *(MZ)*, and fraternal or dizygotic *(DZ)*. The former develop through the splitting, after conception, of one single ovum, fertilized by one single sperm; MZ twins, while not exactly 'identical' on genetic grounds, share a common chromosomal heredity. DZ twins, on the other hand, develop from the fertilization by two sperms of two ova that happen to be simultaneously present in the womb at the same time; they are no more alike than are ordinary siblings as far as heredity is concerned, which means to say that they have 50% of hereditary factors in common. Thus practically all observed (phenotypic)* differences between MZ twins must be due to environment, while for DZ twins heredity itself may account for much of the observed difference. It is clear that these fortuitous games of nature present us with an excellent chance of carrying out genetic experiments; variables which are not influenced by genetic factors at all should find MZ and DZ twins differentiated to about the same degree, as all the differentiation would be due to environment. The more important genetic factors are, the more alike should MZ twins be as compared with DZ twins.

One way of using this approach is to take pairs of twins who have been brought up together, sharing a common environment, so that differences due to heredity will, as it were, stand out in relief. However, this raises one difficult question: are the 'common environments' of MZ and DZ twins in fact equally 'common', or are MZ twins, being more alike, treated in a more similar manner by parents, teachers, and peers? This difficulty is avoided when we consider only MZ twins who have been separated soon after birth. (Ideally one would separate them soon after conception, but this unfortunately is not possible.) The twins share some common antenatal environment, and usually for a few months some common postnatal environment; however, much the greater part of their life is spent in quite distinct and

* *Phenotype* refers to the observed characteristics of an organism; these are due to the interaction of environment and *genotype*, that is the purely genetic causes of the characteristic.

different environments, so that any similarities that appear between them are likely to be due to heredity, and differences to environment. This at least is the rationale of the experiments now to be considered. This rationale is not beyond criticism, and these criticisms will be discussed after we have considered the results.

There are in the literature reports of 122 pairs of MZ twins reared apart after early separation in life. The mean IQ of these twins is 97, which is not so far removed from the population value of 100 as to make them unrepresentative. The mean difference in IQ between members of the various pairs is 6·60 points of IQ (which should be compared with a difference of 4.68 when the same person is tested twice). The correlation between twins works out at .82, but if this is corrected for reliability the value would rise to something like .86 (assuming a reliability of .95) or something like .90, making a more realistic assumption. These figures only have meaning for the calculation of heritabilities if we can assume that twins were assigned to homes on a random basis, that is that the socio-economic status of one twin's home was quite unrelated to that of the other's. This question can be properly pursued only on the 46 pairs of twins tested by Sir Cyril Burt (1966), as he is the only investigator to furnish information on the socio-economic status (SES) of the homes in which the twins were reared. He used six categories—(1) higher professional, (2) lower professional, (3) clerical, (4) skilled, (5) semi-skilled, and (6) unskilled. Seven cases reared in residential institutions had to be omitted, as there is no basis for assigning them to one of the six categories. There was no relationship between the SES of the home assigned to one twin and that of the other (the correlation was .03), and hence none of the correlation between the twins' IQs can be attributed to similarities in their home environments. This may not be true of the other studies using twins brought up in separation, and indeed clearly Burt's study differs in this respect from that of

Newman, Freeman and Holzinger, where there is evidence of some such correlation. However, if this were an important point then Burt's study, showing no correlation, should have the lowest value for the twin IQ intercorrelation; instead it has the highest of the four investigators who have used this technique. We may conclude that while the possibility is not completely excluded that placement of twins in homes has been influenced by the desire to give them similar backgrounds, nevertheless the influence of this factor cannot have been very important.

The twins differ of course with respect to the amount of difference in IQ shown; this would be expected on any measure which is not completely reliable, and which is to some extent influenced by extraneous (environmental) variables. Some critics have taken a particular extreme case of difference (that between Gladys, with an IQ of 92, and Helen, with an IQ of 116) and argued that this difference of 24 points destroys the argument. These twins were separated at the age of eighteen months, and tested at the age of thirty-five years; they had markedly different health histories as children, with Gladys suffering a number of severe illnesses, one being nearly fatal, while Helen enjoyed unusually good health throughout. Curiously enough, the existence of one such high discrepancy does not destroy the case for hereditary influence, but serves to establish it all the more securely. The reason is simple; in science, results can be too good for a model, as well as being not good enough. If your model predicts that 9 out of 10 cases should lie within a given area, then the prediction is falsified equally by finding that only a few cases lie within that area and by finding that all 10 cases lie within that area. Given the mean difference in IQ between MZ twins brought up in separation, and the dispersion of these values, then 1 value in 122 should be as high as 24; if this had been missing, the results would not fit in with prediction.

What are the numerical results of the attempt to measure or estimate heritability? Jensen (1972) has reanalysed the results of the four studies, and comes to the conclusion that the variance can be partitioned in the following way: Heredity = 85%; Environment = 10%; Test error = 5%. Some caution is of course needed in interpreting these figures. In the first place, the twins were not separated at birth, but after a period of six months or so on the average; it is possible that ante-natal and post-natal environmental factors might have contributed to make them more alike than they would otherwise have been. This factor might be expected to raise the hereditary contribution to the variance artificially. (There is evidence to show that this is not so; this will be considered later.) But in the second place, there are factors which might be expected to lower it artificially. Even between *MZ* twins, certain differences occur in the first stages of cell division inside the womb that would cause these twins to be less alike than a simplified account of chromosomal inheritance would suggest. Darlington has (1954) concluded a review of these effects by stating that the total effect of these biological discordances would be sufficient to lead to a gross underestimate of heritability in all twin studies. And in the third place, Price (1950) has shown that there are what he calls 'primary biases' in the prenatal development of *MZ* twins which make for discordances between them during antenatal development.

Taking the factors to which Darlington and Price have drawn attention together, it seems certain that estimates of heritability based on the hypothesis of identical heredity in *MZ* twins are biased, in the sense of attributing to environment effects which in fact are either attributable to heredity, or else belong in a kind of 'no man's land' which is properly neither heredity nor environment in the ordinary meaning of these terms. Taking these three sources of possible error into account, it seems likely that they either balance out, or that the

estimate of heritability arrived at is a slight underestimate. Future work should clearly be much concerned
with establishing in more detail the precise conditions
under which the phenomena in question occur, that is
there should be careful investigation of the environmental variables involved (going beyond the simple estima-
tion of SES) and there should be a careful study of the
birth-history of the foetus. These additional investigations would not materially alter the conclusions arrived
at here, but they would make it possible for us to obtain
a more precise and less biased estimate than is at present
available.

Much more could of course be said about details of
these investigations. However, this is no text book, and
we must remain content with the main results. Let us
merely note one further point: however impressive the
findings, they can only be trusted if similar results are
obtained through the use of other methods of research
logically independent of those used here. It is by such
consensus of methods not subject to similar criticisms
that science builds up a quantitative model on which
reliance can be placed.

We next turn to the study of identical and fraternal
twins. The logic of this approach has already been explained; it rests on the respective observed differences in
IQ between MZ twins, on the one hand, and same-
sexed DZ twins, on the other. Usually these differences
are expressed in terms of correlations, and our model
would predict that these correlations would be markedly
higher for MZ than for DZ twins. The correlation between MZ twins reared together is about .87, that of MZ
twins reared apart is about .75. When we come to DZ
twins, different studies give slightly different values, but
the average is about .50 (for different sexes) to .55 (for
same-sexed DZ twins). Thus there is no doubt about the
existence of the postulated difference. The obtained
values are very close to those expected on theoretical
grounds; assuming assortative mating and partial domi-

nance, the predicted value would be .52 for both types of DZ twins. These studies, then, give us support for the hypothesis of a strong hereditary component in causing IQ differences; can we use them further to help us construct a model of gene action which would enable us to state in more detail just what are the genetic forces which are active in producing these effects?

A determined effort has been made by Jinks and Fulker (1970) to use the most recently developed methods of biometrical genetical analysis for this purpose, using existing data provided by twin research over the years. The methods used are too technical to describe in detail but we may note the outcome of their analyses with some interest. In the first place, they found that 'the inheritance of most of the psychological measures reanalysed conform to a simple model. In view of the pessimism over the possible influence of correlated environments and genotype-environment interaction so often expressed in the psychological literature, it is reassuring to find they are by no means universal phenomena.' In other words, interaction effects, so often suggested as making the apportionment of variance between heredity and environment impossible, were found to be either non-existent or quite unimportant. This is crucial because it removes one possible source of criticism of many previous studies using the older type of methodology, which simply disregarded the possibility of such interaction effects. In the second place, Jinks and Fulker found that between families* environmental factors were unimportant; this too is interesting for the same reason; the older methods, by having to leave out of account such factors because of defective statistical tools available, were open to criticism on theoretical grounds, and now these criticisms can be seen not to have had any

---

* Environmental effects can occur 'within families', that is they are due to differential treatment of children within the same family, or they can occur 'between families', that is they are due to causes that differ from one family to another.

4

great strength. In the third place, it was possible to demonstrate that directional dominance was present in the model (high IQ being dominant over low IQ) as well as assortative mating; thus our model is becoming much more precise than was possible before. And in the fourth place, they showed that the genetic determination of IQ not only involved a number of different genes (which had been established before) but that the number of genes involved was probably about 100. As they point out, 'gene action strongly suggests that IQ has been subject to considerable directional natural selection during man's evolutionary history'. (In these analyses, use was made of data additional to that involving *MZ* and *DZ* twins, but the major analyses relied on these groups.)

Studies along these lines are often criticized on the basis that *MZ* twins may be treated more alike by the environment than are *DZ* twins, and that consequently the assumption of equal environment for the two types of twin pairs is unjustified. There is no doubt that in many cases *MZ* twins are treated more alike, but it must be doubtful if the manner in which this is done is very relevant to intellectual achievement. The objection is probably more applicable to studies of the inheritance of personality variables, and even there empirical evidence has been given to show that the effect is not strong, and may be completely absent. The point will be taken up again in a later chapter. To demonstrate that this is an important source of variation in IQ performance, it would be necessary to show that parents discriminate between *DZ* twins in a manner which can be demonstrated to affect IQ; this has never been done. And seeing (as we shall presently) how difficult it is to change IQ scores even by the most determined efforts to do so, and how small the effects in fact are, it seems extremely unlikely that the small differences which parents may show in their treatment of *DZ* twins (as compared with the differences they might show in their treatment of *MZ* twins) could produce any sizeable effects which would force us to reconsider our estimates of heritability arrived at from these data.

One further result from the Jinks and Fulker study is of particular interest. In addition to analysing IQ performance on several tests they also analysed educational attainments for different types of twins. This is of interest because extreme environmentalists have often maintained that IQ tests are so contaminated by cultural variables that they are in principle no different from tests of educational achievement. If this were true, then similar results should be found for these two kinds of tests; in fact the data show massive differences between the two sets of results. (1) The heritability of Burt's group test of intelligence (which probably resembles the educational tests more than 'culture fair' tests) was found to be .86; that of educational attainments was found to be less than .30. This is a tremendous difference, suggesting that educational attainments are in part due to inherited differences in IQ, and in addition to environmental factors; of these two kinds of determinants, the latter are decisively more important. (2) A simple model of inheritance was found to be adequate for IQ, but not for educational attainments. (3) For educational attainments common family environment was found to be very important, and accentuated by effects of correlated environments; neither of these findings was applicable to the IQ data.

In view of these results, it is not possible to maintain any longer the notion that IQ tests are identical with, or even similar to, tests of educational attainment; the two kinds of tests differ profoundly in so many relevant and important variables that no comparison is possible. Such criticisms of the genetic model are based on assumptions which cannot be supported, and only a blithe disregard of the quantitative evidence can lead anyone to place much value on this suggestion. Whatever IQ tests may measure, it most certainly is not educational attainment, even though IQ does of course play a part in enabling a child to acquire educational attainment.

When we look at the study of kinship correlations we find it is possible to deduce from the genetic model the degree of intercorrelation in IQ to be expected from different degrees of kinship, that is between parents and children, uncles and nephews, first cousins, second cousins, etcetera. Over fifty such studies have been reported in the literature, and summarized by Erlenmeyer-Kimling and Jarvik (1963) in a well-known paper. They concluded from their survey that although the studies had been based on a wide variety of mental tests, and administered under a variety of conditions, by numerous investigators with contrasting views regarding the importance of heredity, yet, 'against this pronounced heterogeneity, which should have clouded the picture, and is reflected by the wide range of correlations, a clearly defined consistency emerges from the data. The composite data are compatible with the polygenic hypothesis which is generally favoured in accounting for inherited differences in mental ability'.*

Figure 4 sets out in detail some of the major findings; expected correlations are indicated by a cross, the observed correlations by a nought. It will be seen that the predicted and the observed values are remarkably close together. The theoretical values are derived assuming assortative mating and partial dominance; as we have seen in our previous discussion, the evidence for both is strong. Note that this is a model which postulates only genetic causes, admitting no environmental causes whatever; a closer correspondence could no doubt be obtained if we constructed our model for expected correlations on the basis of a heritability of .80, rather than 1.00. However this might be, the figures do demonstrate that environmental influences cannot have been very effective in producing the observed values.

* That is, that the inheritance takes place through a number of genes, rather than through a single gene.

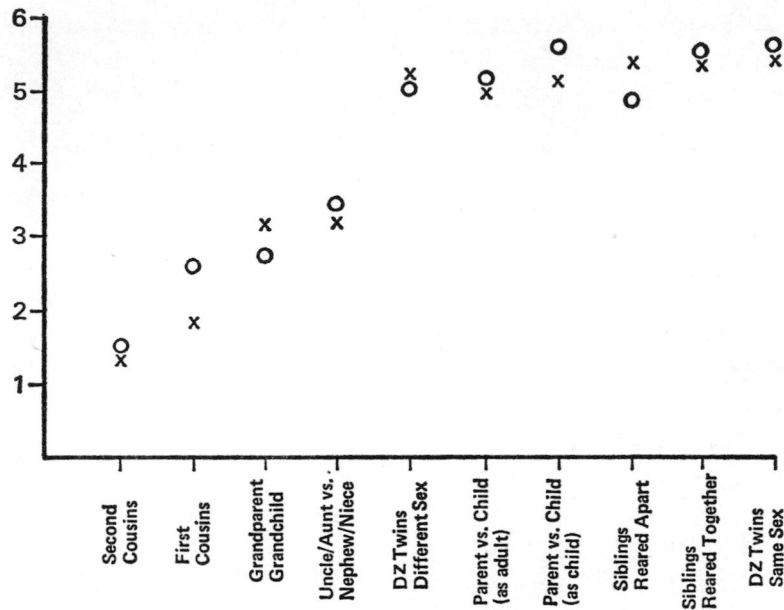

Figure 4   *Familial correlations for intelligence*

Burt (1961) has attempted to further analyse kinship figures collected by himself from London school populations; the results are shown in Table 5 below. He actually used two bases of assessment. The first basis was the child's IQ score: when these scores were considered to be unfair to the child, and not representative of his

Source of Variance	*Per cent*	
1 Genetic:		
Additive	40·5	(47·9)
Assortative mating	19·9	(17·9)
Dominance and Epistasis	16·7	(21·7)
2 Environmental:		
Co-variance of heredity and environment	10·6	(1·4)
Random environmental effects	5·9	(5·8)
3 Unreliability:		
Test error	6·4	(5·3)
Total phenotypic variance:	100·0	100·0

Table 5   *Breakdown of test-score variance*

actual intelligence, the child was retested, often being given several different tests on several occasions. This resulted in an 'adjusted' assessment of the child's IQ. Environmental factors contribute 16·5% to the variance when the raw scores are analysed, and 7·2% when the adjusted scores are analysed. Unreliability of the test contributes some 5% to 6%, and the rest is hereditary variance, amounting to 77% for the raw scores and 88% for the adjusted scores. Analysis of kinship correlations thus gives figures not far removed from those we have found in conjunction with our other methods of investigation.

We find that inbreeding has certain obvious genetic consequences on variables which are strongly determined by heredity and consequently the study of inbred individuals is of great value. Schull and Neel (1965) carried out such a study in Japan after World War II, studying the offspring of marriages of first cousins, first cousins once removed, and second cousins. The parents were carefully matched with a control group of unrelated parents of equal age and socio-economic status. The degree of consanguinity produced by the cousin marriages had the effect of lowering the observed IQ of the inbred children by nearly eight points as compared with the control group. This is a very strong effect when it is realized that first cousin marriages produce children who only have one out of sixteen pairs of genes by common descent; for marriages of first cousins once removed, the children have only one out of thirty-two pairs of genes by common descent. For marriages of second cousins, the number of pairs of genes the children have by common descent is only one in sixty-four. It is noteworthy that the IQ effects were among the clearest and strongest of all those observed; other observations made included physical illness, several anthropometric and dental variables, and school grades. This lowering of IQ after inbreeding is perhaps among the best lines of evidence we have for hereditary control over intelligence, and it definitely suggests that the system involved is multifactorial, not one controlled by a small number of genes. In both these points, the data agree perfectly with those already considered.

The phenomenon of regression is so important that it will be discussed in greater detail again in a later chapter; here let us merely note its existence and meaning, and its relevance to the question of the heritability of intelligence. Figure 5 will serve to introduce the meaning

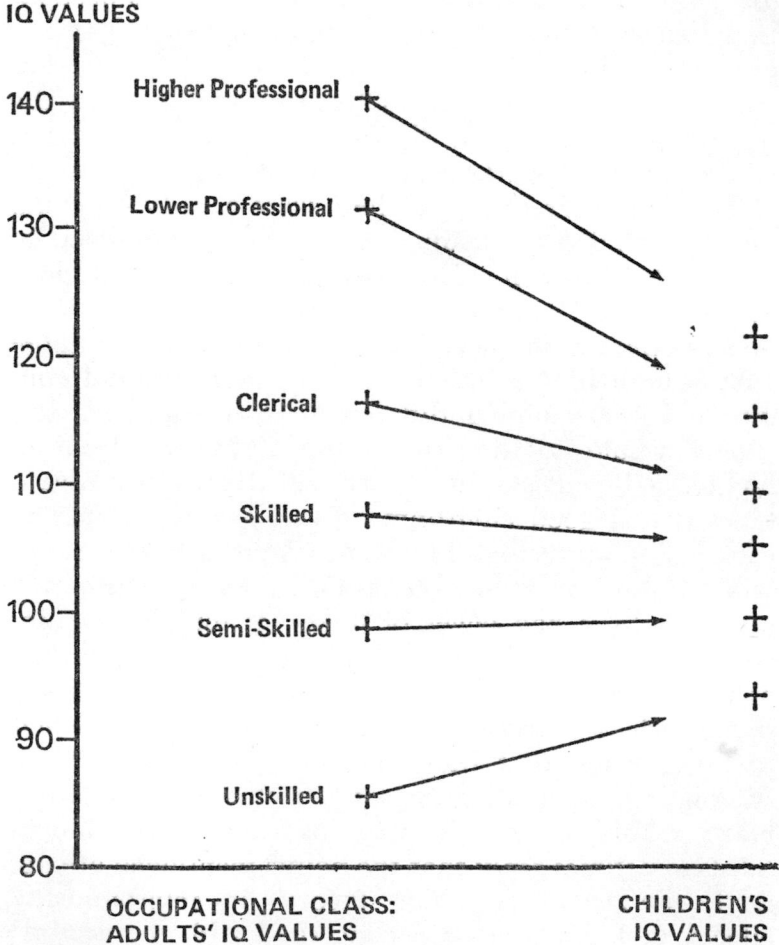

IQ VALUES

OCCUPATIONAL CLASS:
ADULTS' IQ VALUES

CHILDREN'S
IQ VALUES

Figure 5   *Regression to the mean*

of this term. It shows on the left side of the diagram the IQ values obtained by large groups of fathers in various occupational classes, ranging from the upper professional through the lower professional and clerical to the skilled, semi-skilled and unskilled working-class groups. It will be seen that the values are not very different from those given in Table 4: the present data were collected by Burt (1961) in this country. On the right side of the diagram are given the corresponding values for children whose fathers are in these various occupational classes; it will be seen at once that they have regressed towards the general mean of the population (which is of course 100). Children of fathers in the higher professions show the greatest decline; children of the unskilled workers show the greatest increase in score. It almost seems as if there were a deity watching over our affairs, determined to reduce the differences in IQ which at present exist between different occupations and classes.

This effect would certainly not be predicted by those who believe in the importance of environmental conditions in determining differences in intelligence. The theory would predict that the children of well-off, highly intelligent parents, having all the worldly advantages that a good environment can provide, would at least maintain the high intellectual level of their parents, and possibly exceed it—seeing that their parents would normally have had fewer of these advantages (a point which will be proved presently). Conversely, the children of the unskilled, one might have expected, would not show such a marked improvement in IQ over their parents, seeing that these can only provide the least stimulating, least advantageous background as far as material things are concerned. In answer environmentalists sometimes reply that the points mentioned are not really important in their theory of environmental effects, and that they are concerned with more subtle things. Yet socio-economic status has always been one of the environmental factors which could be demonstrated to play a part in affecting IQ, and political changes favoured by environmentalists are usually aimed at increasing equality precisely in such things. In particular, equality of schooling is one of the declared aims of egalitarians. Yet the ability of the higher-ranking parents to buy for example supposedly advantageous types of schooling does not seem to have profited their children; they have not only failed to improve on their parents' IQ, but have grievously fallen behind! And the allegedly poor schooling of the sons of working-class parents has had the effect of lifting their IQs above those of their parents!

On genetic principles, however, the phenomenon is perfectly understandable. Geneticists, in fact, have worked out a formula which enables us to predict quite accurately the outcome of such a study. The formula reads:

$$O = M + h^2_N (P - M)$$

In this formula, O stands for the predicted mean IQ of the offspring, M is the mean intelligence of the population in question (100 in our case) and P stands for the mean parental IQ, that is the average of mother's and father's IQ. The value h^2_N stands for the narrow heritability, that is the purely additive portion of the heritability, excluding dominance and epistasis* effects. According to the Jinks and Fulker (1970) analysis this would be about .71, as we have seen. We will see presently whether the formula gives accurate results: here let us merely note that the formula tells us that the child's intelligence will be a function of his parents' intelligence and of the degree to which intelligence is inherited; the greater the heritability, the less will be the regression to the mean. On the other hand, the more the parents' intelligence deviates from the mean, the greater will be the regression effect.

* Epistasis refers to the masking of the effects of one gene by another.

Jensen (1973) has worked out an actual example, using data published by Terman in his 'gifted children' research to which reference has already been made. The mean IQ of the gifted group was 152; that of their spouses was 125, showing considerable assortative mating. Thus the parental mean IQ would be 138.5; M of course is 100, and as our closest estimate of the narrow heritability we have chosen the value of .71. Substituting these values in the formula, we obtain a prediction of 127 as the expected level of IQ for the offspring of our gifted children, now grown up. The actual figure is 132.7; this discrepancy, however, vanishes when we make a necessary correction. The formula assumes randomized environments, but of course the environment provided by our gifted parents was not average but very superior, and we must correct for this superiority. When this correction is made, the predicted value comes to 132.2, which is very close to the obtained value of 132.7. Thus the formula works astonishingly well, although of course this particular degree of accuracy is not likely to be achieved every time it is used!

Two points are worthy of note. For the purpose of the correction in our predicted figure the actual income of the families concerned was used, as a rough-and-ready measure of socio-economic status, and indirectly of the environmental influences which might affect IQ. As we have seen, such very tangible indices of environmental influence are looked down upon by many environmentalists who call for something more subtle, although they seldom specify just what these subtle factors might be, or how they could be measured. But the calculation outlined above shows that income is sufficiently close to the myriad factors which do affect IQ to make it possible for us to make a perfectly good and accurate prediction of the IQ level of the children of our gifted parents; it is in this type of numerical correspondence between predicted and observed values that the model of hereditary determination seeks its main support. The other point also relates to the quantitative nature of the work. We have used a value of h^2_N which was originally derived from studies of MZ and DZ twins; yet this value enables us to make predictions with respect to the IQ of children who were tested with quite a different type of test, by different investigators, in a different country, and at a different period of time. It would seem unreasonable to deny that the numerical value assigned to the narrow heritability (or the value of .80 assigned to the broad heritability) is more than the psychologist's figment of the imagination, or a 'meaningless' construct which has no scientific standing. Many other examples could be given of such numerical comparisons, but to most readers the demonstrations would not be very meaningful as they require a background in the biometrical genetic system on which they are based. Let us merely note that they do exist, and that any effort to criticize the present model must be able to demonstrate (a) that the predictions are not in fact accurate, and (b) that a better model exists which will account for all the known facts. Simple verbal criticism will not do when a theory or model is based explicitly on a mathematical and experimental basis.

The rationale for the study of orphanage children can be spelt out very simply. If, as environmentalists assert, the variability in intelligence shown by members of a population is due to differences in upbringing and other environmental variations, then it would follow that any very marked restriction in the variability of the environment to which a group of children are exposed would lead to a restriction in the variability of the IQs of these children. In other words, they should all be very much alike with respect to intelligence. Now orphanages to which children are assigned fairly soon after birth, and in which they are kept for many years, would seem to fit this prescription. Conditions are as uniform as human ingenuity can make them, with meals, schooling, free time, holidays, staff, provision of books and other entertainments made as uniform as possible. Obviously uniformity cannot be absolute, but it would be impossible to deny that in an orphanage there is a tremendous restriction in environmental variation, and this ought to show itself in the variability of IQ of the children exposed to these conditions. Is this actually found? It will be seen that we are now concerned with a direct study of environmental influences, where in the previous five sections we have discussed direct genetic experiments: these five sections do not of course exhaust the many applications of genetic experimentation, but they must suffice for the purposes of this book.

Lawrence (1931) compared the variability in IQ of orphanage children and ordinary school children, exposed to all the divergent and multifarious environmental conditions obtaining in the outer world. The average age at which the orphanage children were admitted was around six months; none were taken after they had passed their first birthday. IQ tests were administered to both groups, and the standard deviations calculated as a measure of variability. The orphanage children did in fact show somewhat smaller variability, but the difference was barely significant: in fact, it was smaller than that observed between the boys and the girls in the control group of ordinary school children tested! Even if we regard the observed slight difference as replicable, its size would be in good agreement with the genetic model, which after all allows something like 20% of the variance to be determined by environmental causes; the diminution in variance is less than 20% for the orphanage children.

It may be said, of course, that the uniformity of environmental conditions was very partial, and may not have included important but subtle influences; furthermore, antenatal conditions and post-natal conditions for a short period were not controlled (but see a later section). All this is of course true, but looked at from the point of view of a determined egalitarian the results are nevertheless devastating. If a government were determined to enforce exactly equal pay, equal education, and equal housing on all its citizens, it could hardly do better than the orphanage; if in addition it separated children from their parents to prevent differences in parental knowledge and intelligence from giving some children a better start in life, it could hardly do more than the orphanage. Yet the outcome of all this would hardly be commensurate with the effort; the reduction in human diversity of intelligence would not be more than a trifle, if that. This experiment holds many important truths: it tells us by direct experimental demonstration just what are the present-day limits beyond which we cannot go because nature has drawn definite limits of biological causation.

This truth may be put into diagrammatic form. Jensen (1972) has calculated the shrinkage to be expected in human variability in IQ if all environmental sources of variance were removed; Figure 6 shows the result, assuming a heritability of .80. It will be clear that this result is not very different from that found in Lawrence's orphanage experiment: this suggests that those environmental variables which were made uniform in this study embody most of the sources of variance

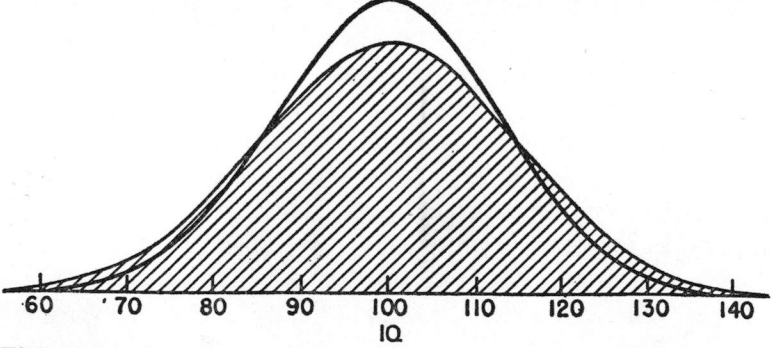

Figure 6 *Actual and hypothetical distribution of IQ in the population*

which are responsible for the non-genetic portion of the total phenotypic variance. This in turn suggests that the hypothetical criticisms of the study given in the preceding paragraph are not in fact justified; the possible 'subtle' environmental effects do not seem to find a place in the mathematical model with which we are concerned, and would require direct experimental proof before they could be taken seriously. Figure 6 deserves close study by all those who hope to change people by changing society; it shows clearly the narrow limits within which such an enterprise must be confined at present, even if we could control all environmental sources of variation.

Foster children separated from their true parents shortly after birth and reared in adoptive homes present us with another important natural phenomenon which enables us to test deductions from our model, and from the environmentalist hypothesis. The model would predict very modest correlations between child's IQ and foster parents' IQ, but correlations between child's IQ and natural parents' IQ only marginally lower than those found when children are brought up by their own parents. Environmentalists would expect little or no correlation between child's IQ and IQ of natural parents, but a considerable correlation between adopted child and adoptive parents.

The actual facts are very clear-cut, and bear out the deductions from our model. On the whole, children separated from their parents early in life, and tested as adolescents or adults, have IQs which correlate with those of their natural parents to almost the same extent as would be the case had they been brought up by their biological parents. The correlations between adopted children and their foster parents for IQ range from zero to .20, demonstrating once again the small influence which environmental changes even of such a far-reaching sort have on the growing child's IQ. All the 'subtle' environmental influences postulated by confirmed environmentalists, which might be acting in the interrelations between foster parents and foster children, do not seem able to destroy the biological reality of inherited IQ level; neither do the more obvious ones of socio-economic status, income, and education. It is interesting to note that even when adoption agencies attempt to place children selectively, by placing children from a high status parent with foster parents of a similar status, there is little increase in the observed correlation between foster child and foster parents. Note also that the correlation between foster child and natural parents increases during the time that the child is actually with the foster parents, until it approaches a final value of .50 or thereabouts by the age of five or six; in other words, the changes in IQ level which are so characteristic of very young children are uncorrelated with environmental factors, but are produced by maturation controlled by genetic factors.

This point is important, and may be supported by an experimental study reported by Ronald Wilson (1972) designed explicitly to investigate this point directly. In this experiment he appraised the mental development of infant twins during the first and second years of their lives. Twins showed high concordance for level of mental development and, what is of particular interest, they also showed concordance for the spurts and lags in development during this period—*MZ* twins more so than *DZ* twins. 'From these results it was inferred that infant mental development was primarily determined by the twins' genetic blueprint and that, except in unusual cases, other factors served mainly a supportive function.' The families in this study ranged from the welfare case to the wealthy professional family, but there was very little relationship between the socio-economic status of the family and the overall level of development during the second year (the correlation was only .2).

Clearly the whole course of development of a child's intellectual capabilities is largely laid down genetically, and even extreme environmental changes, such as transfer from the natural mother to a foster mother, have little power to alter this development.

This is a finding particularly relevant to a widely held hypothesis, the so-called 'critical period theory'. Work with animals has shown that early environmental enrichment or deprivation influences later problem-solving ability, and it has been suggested that in humans too it may be very early family influences (taking place during the first two to twelve months of the baby's life) that determine his later IQ. The evidence considered above would seem to contradict this view, and so does a study specially performed by R. C. Johnson (1963) to investigate this matter. He compared identical twins who had been separated at a mean age of two months with another group of identical twins separated at a mean age of twenty-four months; it would follow from the 'critical period theory' that the twins separated later on should be much more alike with respect to IQ than those separated earlier on. In actual fact, the mean difference between twins separated early in life was 4.7 points; that between twins separated later in life 9.4 points. The similarity in IQ between identical twins was thus *inversely* related to the amount of time spent in a common environment in infancy and early childhood. Johnson, after considering several possible hypotheses to account for the finding, none of which were found to apply to the data, concludes that the 'critical period theory' is not supported by his results. It will be remembered that the possibility of some such specially important early period of life together was the main criticism of the 'twins reared apart' proof for the importance of genetic influences; we can now see that this criticism has no force as far as human subjects are concerned.

Given the existence of foster children, it follows that we should be able to study the influence of environmental variables by directly measuring these in the homes and environments of the foster children concerned;

granted a sufficiently varied batch of children (with respect to IQ) we should then be able to correlate each of our environmental indices with the final IQ of the child, and determine in this way what influence the different variables have had on his intelligence. We would also be able to combine the many different variables and weight them in such a manner as to produce the highest possible multiple correlation; this would tell us the total possible effect of all our environmental variables taken together. Selective placement is of course a problem; we would wish to be sure that this had not taken place, as otherwise we would overestimate the influence of environment. In the famous study by Burks there appears to have been very little selectivity of placement, and in any case any errors in this respect would work against the genetic hypothesis, and would therefore tend to invalidate our model.

Burks (1928) was very careful in rating the adoptive homes in as detailed and fine-grained a manner as possible, spending 4 to 8 hours of individual investigation on each home. She included intelligence measures on the adopting parent as part of the child's environment, as well as such things as time spent on helping the children do their home work, amount of time spent reading to the children, number of books in the home, education of both parents, parents' vocabulary, a culture index, and many others. She calculated a multiple correlation between child's IQ and all the variables optimally weighted, corrected this for unreliability of measurement, and came up with the figure of 18% of the total variance being explainable by the environmental factors considered. This agrees well with our model, which allows something like 20% for environmental contribution to variance. It should be noted that Burks' method of analysis almost certainly overestimates the value for environment because the formula capitalized on chance errors. Leahy (1935) attempted a similar study, but simply summed up all the environmental indices he used; he only found 4% of the variance accounted for by environmental variables. However, the indices he used were inferior to those used by Burks.

The Burks study is so careful, both in its experimental and its statistical expertise, that although it was done as long ago as 1928 it still remains unsurpassed. The results fit in very nicely with our model, of course; critics usually suggest that some mysterious and unspecified factors in the environment may have been missed out. This is of course possible, but is not a testable hypothesis; unless a factor is specified and shown to be measurable, there is little that can be done with such a view. Unspecified and untestable variables, introduced into the argument merely to salvage an untenable position, are not part and parcel of the traditional scientific approach. If any particular variable which has not been measured by Burks, and which is not correlated with any of the variables used by Burks, can be shown to correlate with IQ, then the study could be repeated with that variable included. Until this is done the criticism is somewhat empty and unhelpful.

The most convincing support of the environmental hypothesis would of course be a direct demonstration that IQ can be raised by environmental manipulation along traditional lines, by improvements in diet, home circumstances, education, etcetera. Many such studies have been reported, and many successes have been claimed, but there are certain persistent errors of design and statistical treatment which beset many of these studies. The two main errors are (1) neglect of regression effects, and (2) 'teaching the test'. The second of these is the more obvious. Essentially what we are trying to do in manipulating the environment is to change a person's *general ability*; we are not concerned with merely changing his ability to pass a particular IQ test at a higher level than before. We could achieve the latter aim very simply by telling him all the answers, and making him learn them by heart; this would assuredly give him a higher

score, but would hardly improve his IQ in any meaningful sense! Yet in too many studies this is precisely what
has been done, although not perhaps in quite so direct
and obvious a manner; more usually a more roundabout
method of 'teaching the test' is adopted. Unless we can
be sure there has been no attempt to do this, the results
as reported are not worth very much.

The problem of statistical regression gives rise to an
obvious source of error, but it is curious how many investigators have failed to take it into account. Let me
give an example to make clear what is involved. Suppose
that we select a target population for our experiment,
for example a group of children having IQs below 80.
Now a child's IQ is of course open to errors of measurement; on repeated testings it may vary by a few points
up or down the scale. Suppose that on a given day we
test 10 children all of whose IQs are exactly 80: because
of day-to-day variation the measured IQs might come
out at 71, 74, 75, 76, 78, 82, 84, 85, 86, 89. For our
experiment we would select the first 5 subjects, rejecting
the other 5; our experimental group would have a
mean IQ as measured of 75. We might use the other 5
children as a control group, that is a group not given
whatever educational technique we are using for the
experiment; this group would have a mean IQ of 85.
Now we might retest both groups at the conclusion of
our experiment; lo and behold, the experimental group
has miraculously improved by 5 points, and now scores
80 on the average, while the control group has declined
in IQ, and now scores 5 points lower than before, namely
80 as well. Thus we would conclude that our technique
of educational intervention has produced a difference of
10 points of IQ, which is well worth having; yet in truth
the whole thing is an artefact only, produced by the
unreliability of the measuring instrument. This is a
grossly oversimplified example, of course, but many
investigators have fallen into this trap, and their claims
are still often quoted by convinced environmentalists as

proof of the tremendous impetus which environmental enrichment can have on the IQ of deprived children.

These are of course not the only faults which can be found in many of these studies; many others appear from time to time, and it is decidedly unsafe to accept any reports of this kind without the closest technical scrutiny—there are few areas in psychology which contain so many difficulties and dangers into which the unwary can fall. Instead of examining all the many studies which can be found in the literature I will rather look at one of the most recent and most widely quoted experiments of this kind, sometimes called 'the miracle in Milwaukee'.

The 'Milwaukee Project', to give it its proper title, was originated and supervised by Professor Rick Heber of the University of Wisconsin; it is still in progress. The investigators intervened in the environments of disadvantaged infants and the claim has been made that they succeeded in raising their IQ levels by over 30 points, that is from dull normal to superior! Ellis Page (1972) has subjected this claim to a devastating criticism, and because this study has been so widely cited as proof of the inaccuracy of genetic claims, it may be useful to mention these criticisms briefly. The design of the study is relatively simple. Infants in slum environments, and coming from low IQ mothers, were selected as likely to develop into low IQ adolescents; they were divided into an experimental and a control group, and the experimental group was then subjected to an enriched life curriculum—they were taken early in the morning from their slum homes, driven to the project site, and spent the whole day in the company of a specially trained intelligent mother-substitute who had nothing else to do but play with the infant and give him the sort of enriched background and environment which he would not have had in his real home. In addition, the mothers of the experimental children received home help and training, advice and guidance. Altogether the project made up as many of the environmental deficiencies under which these families laboured as was possible. The result obtained after four or five years of this work has already been described; there was a difference of some 30 points between experimental and control children.

Page has made three major criticisms—in addition to the heart-felt complaint that in spite of the large claims made for this study, and the welcome the results received from the public, the study has never been written up and published in any scientific journal! The first criticism relates to the random assignment of subjects to groups; clearly the design of the experiment demands that the experimental and control children should have been assigned to their respective groups in a manner not dependent on their background, abilities, etcetera. Yet it is quite clear that allocation was not random in any strict sense, and furthermore, in a variable such as height Page could demonstrate that the two groups differed so profoundly at the age of two years that the difference could not possibly have occurred by chance. (There were also differences in weight, chest circumference, heart rate, blood pressure, head circumference, age of tooth eruption, etcetera.) The only alternative to the hypothesis of non-random selection would be one of treatment effects—in the sense that the treatment which the experimental group received had stunted their growth to a degree which makes the claimed IQ change appear very small beer!

The second criticism made by Page relates to 'teaching the test'; this is explicitly admitted in an unpublished paper by Heber. He says: 'Our experimental infants have had training on items fortuitously included in the curriculum which are sampled by the tests. . . . To some extent, infant intelligence tests must contain material which approximates to material used in preschool curricula, primarily because of the limited variety of material for this age.' The reasons Heber gives are in fact well taken; the difficulties he describes are very real. Nevertheless, the fact that items on which the children were explicitly trained overlap markedly with items in the test make the claimed IQ increment meaningless. Clearly one will have to wait until the children are grown up before making any claims for definitive improvement in their IQ; even if this were to be found, of course, the question of random assignment would still remain.

Page has a third criticism to make, namely that the programme of stimulation which was used by Heber is not in fact described in any detail; this would make it impossible to replicate the work. This criticism, unlike the others, is not fatal to the project; the omission can of course easily be rectified. Page sums up his view of the project as follows:

> The Milwaukee Project, then, is here viewed as deficient on three counts: biased selection of treatment groups; contamination of criterion tests; and failure to specify the treatments. Any one of these would largely invalidate a study. Together, they destroy it. Further serious questions have emerged about the availability of technical information for the scientific community. Yet the Milwaukee Project may be one of the most widely publicized studies in educational history. Its 'results' are known to millions. And it may exert an influence over national policy.
>
> How can we set the record straight? A miracle makes exciting news. The failure of a miracle is dull stuff, unappealing to the press. Will society continue to believe the Project's central message (the great plasticity of general intelligence)? Possibly so: it seems to support a cherished philosophical dream.
>
> Still more important, how can our profession better protect the public it serves from this *kind* of event and its consequences? This question deserves serious, responsible study by the broad communities that have an interest and stake in research and development in education.

This is not the only project where questions such as these could be raised. There appears to exist a need for people to believe in the infinite plasticity of human mental ability, and any project that seems to favour this belief is uncritically accepted, widely praised, and becomes universally known even when in fact no published record exists of its findings! On the other hand, studies which go against the feeling of the times, however well designed and executed, are criticized on irrelevant grounds and dismissed from consideration. When educational policy depends on knowledge of the facts concerning children's intelligence, and the variables affecting it, this is a very serious situation; we may recall the sad series of events in Soviet Russia when political preconceptions elevated the anti-scientific ideas of Lysenko to a position of social dogma, and not only practically ruined Soviet biology but also had a deleterious effect on agriculture generally.

When we turn to the study of mental defect we find it may arise from two quite distinct causes (apart from birth injuries and similar factors). (1) Mild defect (feeble-mindedness), characterized by IQs between 50 and 75, is usually simply the tail end of the normal distribution of intelligence. The siblings of such children have IQs which have regressed to the population mean, that is are about half-way between that of the feeble-minded child and normality—somewhere around the 80 mark. Furthermore, siblings of feeble-minded children show a distribution of scores which could be predicted from our genetic model. (2) Severe defect (imbecility), characterized by IQs below 50, is usually due to single recessive genes, or mutant genes, whose effects are so strong as to override completely all other genetic or environmental effects involved in intelligence. Siblings of such children have a much higher level of intelligence than do siblings of feeble-minded children. Furthermore, their parents come from all social classes almost randomly, while the parents of feeble-minded children tend to come predominantly from the lower working class. All this agrees well with our model, and suggests that while feeble-minded children form part and parcel of the general population to which this model is applicable, imbeciles do not.

Our model would suggest that, taking only the feeble-minded (called 'retardates' in the USA), such retardates would be found mainly in families containing other retardates. Elizabeth and Sheldon Reed (1965) have found this to be so. They studied the IQ test results of 1,450 retardates and their 1,951 surviving offspring. When a retardate in the sample married a spouse who was also feeble-minded, 39% of the offspring were also feeble-minded (here defined as having an IQ below 70). Note, however, that the very poor intellectual environment provided by these matings did not prevent 24% of their children from obtaining IQs of 90 or above, a clear illustration of regression to the mean. When a female retardate married a normal spouse, 19% of their children had IQs below 70, 31% ranged from 70 to 89, and 50% had normal or better intelligence. Two normal parents, neither of whom had a feeble-minded sibling, only had a chance of having a feeble-minded offspring of 0.53; this risk rose five-fold when one of the parents had a feeble-minded sibling. Clearly, feeble-mindedness 'runs in the family', but is subject to regression, as expected.

Forty-three per cent of the retardates never reproduced; this may be the explanation why research has never discovered any support for the intriguing notion that our national intelligence is declining. This view was based on the well-documented fact that when you measure children's IQs in school, the duller ones tend to come from larger families; there is a correlation of —.25 or thereabouts between IQ and size of family. If the duller members of society reproduced themselves in larger numbers than the brighter ones, then a fall in average IQ of 1 point per decade would be expected, but this has never been found in replicated studies with intervals of 10 or 20 years. But of course the design of the original experiment, which begins with children in school, leaves out of account families having no children at all; if the dullest families have no children, then the conclusion of a decline in national intelligence does not follow—indeed, there might be a mild increase. The fact that of the retardates in the Reeds' study 43% never had any children at all suggests that the overall picture is not likely to depart much from a 'steady state'; if there are any changes in the national mean IQ, these would be too small to measure.

We have now examined in some detail a variety of proofs of the accuracy of the genetic model of intelligence; we must now turn to some objections which have been made against it. It is much more difficult to state the other side, as it were, than it was to state the case *for* the genetic model; some of the reasons for this disproportion have already been discussed. The model is precise, quantitative, and experimental; the various propositions, deductions, and facts which it contains are interconnected very closely, and present a formidable array of evidence in favour of the model. The criticisms lack most of these features; they are semantic, unconnected, and non-factual. There is no attempt to construct an alternative model which would account for the known facts; rather, objections are voiced against one or the other proof, without attention being given to the fact that many other proofs point to the same conclusion without encountering this particular objection. Thus for instance it is sometimes said that the case for the model rests on twin data, and that twins are so unusual a phenomenon, and so unrepresentative of the population, that data derived from them are irrelevant. Such a criticism is in itself meaningless; it is not made clear why being a twin should excommunicate one from the rest of the living! But more particularly, such criticisms leave out of account the fact that we can take the estimate of heritability derived from twin studies, and use it to predict the regressed IQ of the offspring of Terman's 'gifted children', with very great accuracy. If twins were really *ultra vires* as far as genetic analysis is concerned, then estimates of heritability derived from this source should not agree with estimates derived from other sources, nor should they give accurate predictions in connection with other types of data. But both these possibilities have been shown to be factually verified; consequently the criticism cannot be taken very seriously. It isolates one of many interconnected facts and proofs, where proper criticism should take into account all the facts. This failure to take into account all the facts pervades the criticism of the genetic model, and makes it unsatisfactory from the scientific point of view.

Our discussion here will concentrate on two major criticisms which have often been made by serious students, and which deserve a serious answer. The first of these centres on malnutrition, and in particular malnutrition occurring during what is sometimes called the 'critical period', that is, during the ante-natal period and shortly after birth. (We have already looked at some empirical work disproving this notion of a 'critical period'.)

This hypothesis states that developing organ systems are most vulnerable at the period of maximum growth. Interruption of development at a critical period is likely to be irreversible or, at the least, subsequent development is likely to be retarded; hence prenatal and early postnatal exposure to conditions of famine would, in terms of this hypothesis, have the most severe effects on the intelligence of the child. An excellent study by Stein and his co-workers (1972) is available which submits this hypothesis to searching investigation. Cohorts of children born at varying periods after the famine imposed by the Germans on certain regions of Holland during the war (as retribution for the participation of Dutch workers in the battle following the Arnhem land-ing of British paratroops) were compared with children born during the same time in other parts of Holland not exposed to famine conditions. (At their lowest point the official food rations in the famine areas fell to 450 calories, which is a quarter of the minimum standard. Death rates rose sharply, and many deaths were certified as being due to starvation.)

The investigators used three dependent variables: severe mental retardation, mild mental retardation, and IQ scores; the independent variable, of course, was exposure to famine. The study population comprised 125,000 males who had been born in the selected famine and control cities during the 3-year period, 1 January 1944 to 31 December 1946, and who were inducted into the army at about 19 years of age. The following findings were reported: (1) 'The frequency of severe mental retardation among survivors of the birth cohorts is related neither to conception nor to birth during the famine.' (2) The frequency of slight mental retardation equally fails to be related to experimental famine conditions. (3) With respect to the IQ test used (Raven's *Matrices*) 'once more we failed to find an association with the period of famine'. These results, as the authors indicate, 'point either to a high order of protection afforded the foetus *in utero*, or to great resilience of the foetus in the face of nutritional insult, or to both'.

These findings are sufficiently clear-cut to disprove the hypothesis of 'critical growth', as far as the influence of malnutrition on intelligence is concerned. Of course these results should not be taken too far; as the authors point out, 'the results should not be generalized to the effects of chronic malnutrition with a different set of dietary deficiencies such as often occurs in developing countries, nor to nutritional insult in postnatal life'. This is true, although it must be said that if extreme degrees of malnutrition during the most vulnerable period of the child's life have absolutely no effect on his intelligence, then anyone asserting the influence of lesser degrees of malnutrition during less vulnerable periods must be prepared to produce very direct and incontrovertible evidence, ruling out all other possibilities, before much credence can be given to his beliefs. Some such evidence exists for developing countries, but it does not exist as far as such countries as the UK, the USSR, the USA or the European continent are concerned.

The second major criticism relates to the well-demonstrated effects on learning behaviour in animals of early sensory or social deprivation. Hunt has used these experiments in support of the view that such early deprivations may also be at the basis of learning and IQ defects in human children. Such a conclusion seems unwarranted; the facts are not in dispute, but it is very doubtful if they are in any way relevant to the problem of human intelligence, at least in so far as Western countries or the 'Eastern Block' countries are concerned. (Too little is known about the 'developing' countries in this respect to say very much about them.) Deprivations, either perceptual or social, have to be pretty severe in order to have marked and lasting effects on animals; such deprivations are by no means characteristic of 'deprived' children. Sensory and social deprivation is entirely missing, for instance, in the dull members of big city gangs. The contrary, if anything, would be true; these boys and girls have far more stimulation than a typical middle-class, introverted, school-attending boy or girl. Unless direct proof can be given of the notion that sensory or social deprivation is a causal factor in low IQ in any except a few very rare and unusual 'Kaspar Hauser' cases,* we must conclude that these animal experiments, while interesting in their own right, are quite irrelevant to the problems we are now considering.

* Kaspar Hauser is the name of a child reared in complete isolation in a dark cellar.

Possibly more relevant are cross-cultural studies, for example, those of Eskimo intelligence. Unfortunately the concept of 'deprivation' is not usually well defined, but one would surely have to conclude that Eskimo children, brought up in an environment startlingly un-differentiated with respect to many stimuli we take for granted, under economic conditions which are ex-tremely poor, and with considerable family instability and insecurity, would be unlikely to be able to compete, even on 'culture fair' tests, with white children brought up under ordinary Western conditions. Yet these Eskimos, living in the icy wastes far above the Arctic Circle, score at or above white Canadian norms on the Progressive Matrices. They score much higher than Jamaican or American negroes, although these are brought up under conditions much more closely resem-bling those of the white groups in question, and with a much better supply of environmental and social stimuli. Interestingly enough, Eskimos living under the most primitive conditions did better on the tests than those who lived in closer contact with whites and had be-come acculturated. (These studies may also serve as additional proof that the often repeated criticism that IQ tests are made by white, middle-class psychologists for white, middle-class children, and are unfair to child-ren not belonging to this charmed circle, is mistaken. Certainly the makers of the tests employed in these studies did not have Eskimo children in mind when they constructed their test items!)

It is of course not impossible to advance *ad hoc* environmentalistic explanations of these startling find-ings, but these are without proof and do not have any bearing on the 'deprivation' hypothesis. The point to note is that if social and sensory deprivation, or other environmental deprivation factors, are postulated to account for IQ deficits in white working-class or coloured populations, then the logic of the explanation requires absolutely that a severely deprived group, such as the

Eskimos, should show evidence of IQ deficit; the fact is that they do not.

Thus the 'deprivation' hypothesis limps on two feet; the evidence from animal work which is brought out to support it is true but irrelevant, and the evidence from human children does not support it, but rather goes counter to it. Much further work will clearly be needed before the hypothesis can even be put into a form which will be properly testable; at the moment its proponents are far from agreed precisely on what constitutes 'deprivation' and how this hypothetical entity can be measured. Furthermore, it is of course not only necessary to demonstrate that deprivation (however defined and measured) affects IQ; it is required to show that the effects go beyond the 'reaction range' of environmental influences.

This notion of the 'reaction range' is a very important one in any consideration of the genetic paradigm, and the criticisms levelled against it. Briefly, what this term means is this. Given our paradigm, we can deduce from it with considerable precision the limits within which environmental effects must lie. If a particular treatment should be found to increase the IQ level of a given group of underprivileged children, say, by 50 points, and if this study should be found to be experimentally and statistically adequate (that is not capitalize on 'teaching the test', not admit regression effects as part of the improvement score, and not commit the error of failing to assign experimental and control children at random to their respective groups) then it could be said that this was well beyond the range of environmentally produced variation which the genetic model could accommodate. Should such a study, however, produce an increase of 15 points, then it could be shown that the 20% of the variance allocated to environmental factors could well have produced this effect, and that consequently the model did not have to be abandoned. In other words, having assigned environment a certain quantitatively defined role within our model, we can calculate the limits of that role; these limits are defined by the 'reaction range', that is the range of IQ values which is permissible without violating the quantitative terms of the theory.

I have carefully gone through the published accounts of IQ change, due to environmental manipulation, in which both experimental and statistical rules were not violated to such an extent as to make the results inadmissible; I have not found one such study which showed effects outside the 'reaction range' as above defined. Thus there is nothing in the literature to suggest that changes produced by environmental manipulation cannot be accounted for within the bounds of the 20% of total variance assigned to environment. Critics who wish to impugn this conclusion must do more than point to studies which show a change in IQ after environmental conditions had been improved; they must show that this improvement goes beyond the bounds set by the model. Without such demonstration no example can have any value in dethroning the present model.

It would be wrong to end this chapter with the impression that all that needed to be known was in fact known with respect to the inheritance of intelligence. This would be far from the truth. Our model is a good, reasonable, acceptable work-horse; it has replaced the tin-Lizzie which was the original simple classical 'heritability' model by incorporating dominance, epistasis and assortative mating effects, and by giving quantitative estimates of the strength of within-family and between-families effects, and of interaction between heredity and environment. But much remains to be discovered. To take the simplest things first, we know very little about the inheritance of the 'primary abilities' which make up such an important part of the hierarchical model of intelligence. We do know that these too (or at least some of them) owe their variability to genetic causes in part, though of course not altogether, but the data available at present are too fragmentary to say much more than this. This must be an important area for future study. We have analysed *g* into components of mental speed, persistence, and error checking, but nothing at all is known about the degree to which these mechanisms are under genetic control. This is a serious default which should be remedied at the first opportunity. We know that men and women, although usually scoring roughly equal on IQ tests, differ in two ways: men do better than women on numerical and spatial tests, women do better than men on verbal and memory tests. We also know that there is a tendency for male samples to have greater variance, that is, there are more very bright and very dull males, but there are more average bright females. Are these effects under genetic control, or are they entirely produced by environmental causes? We do not know.

There are many other questions that arise, to which we do not know the answer. However, we do know now how to set about finding the answer, and it is to be hoped that within the next few years work in this field will expand and settle some at least of these important questions.

CHAPTER 4

Intelligence and social class

The facts outlined in the preceding two chapters lead to many interesting and socially important consequences; unfortunately genetically uninformed critics have often come to quite wrong conclusions from these premises. One very frequent confusion, which is relevant to the problem of the relationship between intelligence and social class, arises from the erroneous notion that 'like begets like'; this easily leads to the sort of wrong picture of heredity which is illustrated in Figure 7. Here let us assume that a socially important trait, intelligence let us say, has been split into five main groups of scores, and that we have sixty-four parents (taking the average IQ of *both* parents) and their sixty-four children. The IQ scores among the group of parents and the group of children are of this kind: four have scores above 123 or below 77 each, sixteen each have scores between 108 and 122, or between 78 and 92, and the remaining twenty-four have scores between 93 and 107. If like begets like, then the parents with the scores below 78 would beget the children with scores below 78, while those with scores above 122 would beget the children with scores above 122. Thus, if there were marked differences in IQ between working-class and middle-class parents, say, then these inequalities would be perpetuated *ad infinitum*, and a hereditary sort of serfdom would be inevitable. This is the sort of picture which many people have of the genetic model. In actual fact the picture is quite erroneous; nothing even remotely like it happens. We have already seen in our discussion of regression that like does not beget like in this purely mechanical fashion; genes segregate and recombine and produce an entirely different picture.

The sort of thing that does happen can be illustrated by means of an artificial example which is much simpler than the actual model we are using, but these simplifications do not affect the main points which I am trying to bring out. Let us then assume that intelligence is determined by two pairs of genes, the effects of which are of the same magnitude; let us also assume that these effects are additive. We can then construct a diagram (Figure 8) which also starts out with a parental generation showing the same distribution as in Figure 7; however, while

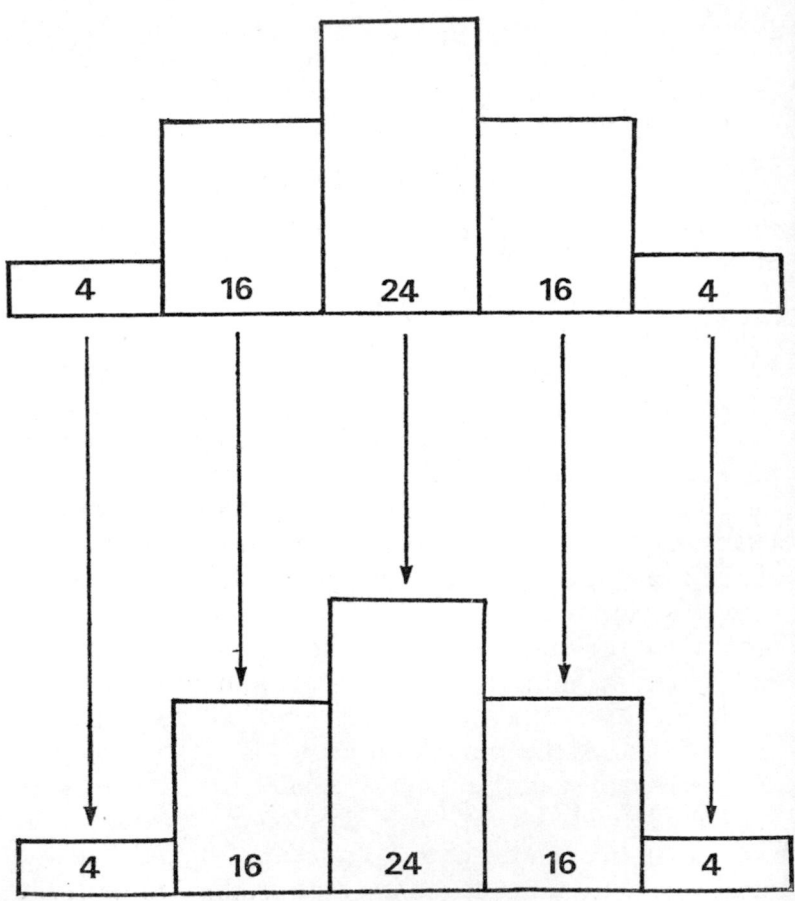

Figure 7   *Erroneous picture of inheritance of IQ*

the distribution of children's IQ is also the same as in Figure 7, the *connections* between the parental and the filial generation have shown a considerable change. Where previously, in the false picture of genetic determination, the lines connecting the two generations lead straight from parent to child, it can be seen that in the proper picture these lines are crossed, so that there are fewer direct connections than indirect ones. Of the four children with low IQs, only one has parents with equally low IQ; the other three have parents with higher IQ.

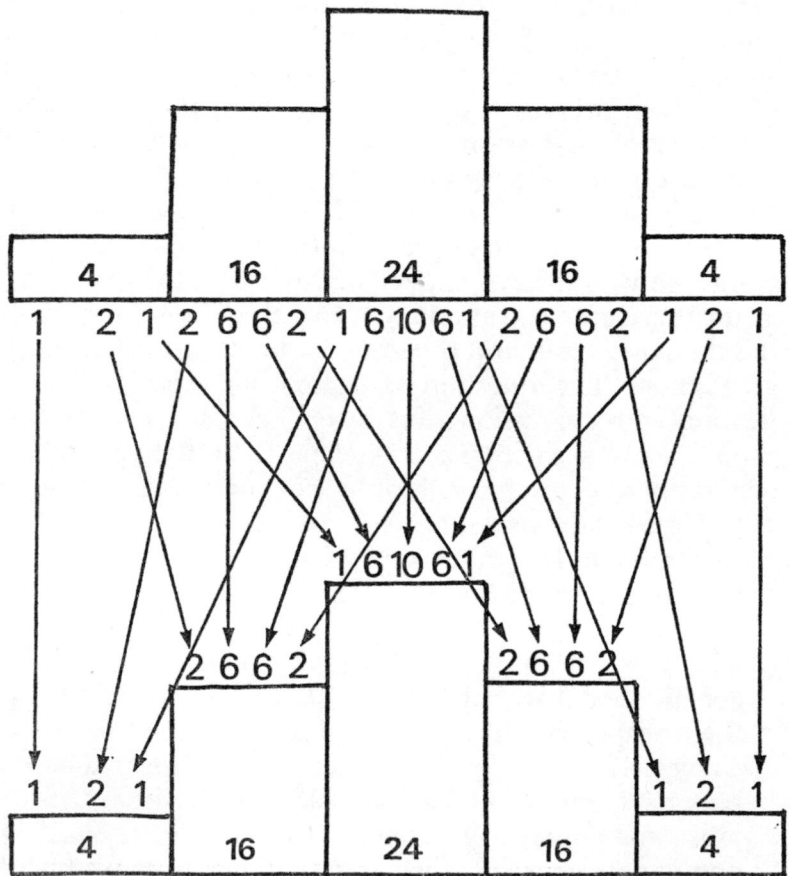

Figure 8 *Correct picture of inheritance of IQ*

Similarly, of the children with high IQs, only one has parents with an equally high IQ; the other three have parents with lower IQs. Of the average IQ children, ten have average parents; one has very bright and one has very dull parents; while six have medium bright and six have medium dull parents. Conversely, if we start out with the parents, we find (very much as indicated in our Figure 5 on page 103) that regression had taken place. The very bright parents only have one bright child between them; the others are average or moderately bright. Similarly, for the very dull parents, there is only one very dull child; the others are either average or moderately dull. Thus of the eight children produced by very bright or very dull parents, two are average! The details of the diagram will speak for themselves; they illustrate the facts of regression better than any commentary. Random gene segregation and recombination is responsible for these effects, and there is no mystery at all about them; from the genetic point of view this is exactly what one would have expected. Note also that the diagram represents a stationary state; the filial generation has the same mean and the same variance as the parental generation. The question is sometimes raised in connection with regression that surely the filial generation should show a shrinking of variance, until after a few generations all children would be born with average IQs. Figure 8 shows why this is not so.

As Professor Li (1971) says (from whose paper these diagrams have been borrowed):

> The most important single phenomenon of the genetic model is that for any given class of parents, their offspring will be scattered into various classes; conversely, for any given class of offspring, their parents come from various classes. Environmentalists sometimes misunderstand the implications of population genetics, thinking that heredity would imply 'like class begets like class'. Probably the op-

posite is true. Only very strong social and environmental forces can perpetuate an artificial class;
heredity does not. From this point of view, social
forces are more conservative than hereditary ones.

The facts of segregation and recombination lead to
one of the most important properties of genetic material,
namely its Markov property (called after the Russian
mathematician who investigated these factors). This
may be put, in simple terms, as follows: 'The properties
of an individual depend upon the state in which he finds
himself and not upon the state from which he is derived.'
In other words, 'A man's a man for all that', as Burns
put it. His IQ determines his mental ability, regardless
of the IQ of his parents; the average IQ persons in our
diagram are exactly equal regardless of the fact that some
have very bright parents, others very dull ones, and still
others average ones. From the point of conception on-
wards, all this is genetically irrelevant. Furthermore,
from the point of view of their progeny, all the people
with genetically identical IQ are equivalent, regardless
of the status of their own parents. Thus if we start out
with a particularly bright father and mother, and trace
the fate of their children, and their children's children,
then we find regression to the mean in each generation,
until after between six and eight generations there is no
predictive value in knowing the IQs of the original
parents at all. As Li puts it:

two individuals of the same family but several
generations apart are practically unrelated in their
genetic constitution. Given the genotype of one, we
could not estimate the genotype of the other any
more accurately than we can for an unrelated random individual. The hereditary forces in shaping
up an individual are essentially of an immediate
nature and have no long-lasting significance. . . . It
is the social forces (created by man) that tend to
protect and maintain a certain class. The genetical

forces (created by nature) have no such social prejudices; they obey the laws of probabilities without memory of the past.

As we have seen, IQ is not determined genetically 100%, and consequently environmental forces present a small obstacle to this levelling process. An example of this has been given in Chapter 3 in our discussion of the regression effects in relation to the offspring of Terman's 'gifted children', where a small correction had to be made in the simple regression formula to take into account the fact that the 'gifted children' provided a better-than-average environment for their children. But the force of this environmental determinant is relatively small; much smaller than high IQ parents would like. To see their children regress to mediocrity, and to be unable to do anything about it, is one of the tragedies of life. It mirrors the tragedy of the egalitarian whose best-laid plans of enrichment of education fail because of the simple laws of heredity. Just as the children of the very bright are bright, but not anything like as bright as their parents, so the children of the very dull are dull, though nothing like as dull as their parents. These are facts which may be applauded or cursed, but which remain facts nevertheless. How do they affect the eternally fascinating problem of social class?

Most people tend to think of social class in terms of Figure 7, that is, they feel that if intelligence is determined by genetic factors to a very marked extent, then the existing class system is likely (with a few exceptions) to become a caste system. If, as they assume, like begets like, and if members of the working class differ from the middle class on the average by some 30 points, then their children, and their children's children, are also doomed to remain in a subordinate position. Yet, as we have seen, exactly the opposite is true. Heredity redistributes the genes which make for superior achievement, high intelligence and great ability, and makes sure that within a few generations none of the existing boundaries between classes shall remain. If anything it is social structure which makes social divisions permanent, acting as the conservative agent; it is heredity which breaks the mould, making for disruption, change and revolution. Wherever there is anything approaching equality of opportunity, universal education, and advancement on merit, there we would expect to find a constant high degree of social mobility, with the children of the middle-class parents regressing to the mean and moving downward, and the children of working-class parents regressing to the mean and moving upward. These, of course, are expectations which apply to the average; within a single family, however, individual children may be quite unlike each other, and again we would expect the bright to swim, the dull to sink. These are expectations based on genetic hypotheses; if intelligence were determined only by environmentalistic causes, then of course classes would indeed calcify into castes, and a permanent serfdom of the dull would be established, without hope of redress. What are the facts?

Consider Table 6 which condenses information given

Categories used: Fathers	N	Percentage allocation of children within fathers' status groups according to category of child						
		Status category: Subjects						
		1	2	3	4	5	6	7
1 Professional and high administrative	129	**39**	15	20	6	14	5	2
2 Managerial and executive	150	11	**27**	23	12	21	5	2
3 Inspectional, supervisory: higher grade	345	4	10	**19**	19	36	7	6
4 Inspectional, supervisory	518	2	4	11	**21**	43	12	6
5 Skilled manual; routine grades of non-manual	1,510	11	2	8	12	**47**	17	12
6 Semi-skilled manual	458	0	1	4	9	39	**31**	16
7 Unskilled manual	387	0	1	4	8	36	24	**27**
N	3,497	103	159	330	459	1,429	593	424

Table 6 *Extent to which sons have the same socio-economic status as their fathers*

by D. V. Glass (1954) in his book, *Social Mobility in Britain*. There are seven categories used to describe social status, ranging from 'Professional and high administrative' to 'Unskilled manual'; along the horizontal axis are given the status categories of the subjects of the enquiry, while along the vertical axis are given those of their fathers. Also given are the numbers of subject in each row and column. The table tells us that of fathers who were in category 1, the sons achieved the same category only in 39% of all cases; 15% fell into category 2, 20% in category 3, and so on down to 2% who fell into category 7. Similarly, fathers who fell into the lowest category had sons of whom 27% also fell into the lowest category; 24% fell into category 6, 36% into category 5, and so on. There is thus a clear regression towards the mean, very much as predicted. Of the sons born to fathers in the top category, 61% have regressed and belong in a category lower than their fathers, while of the sons born to fathers in the bottom category, 73% have regressed and belong in a category higher than their fathers. These data of course relate to a time which is in the past; such data can never give a picture of the contemporary scene, but it should be remembered that when these 'children' were brought up nothing like our degree of equality in education had been achieved, and class lines were much more rigid than they are now. Furthermore, it should be remembered that this study was done in England; social mobility was almost certainly greater in such countries as the USA. We must conclude that there is no evidence here for a 'caste' system of predetermined excellence or mediocrity; there is constant large-scale change from one generation to the next, very much as we had expected on the genetic hypothesis. These data are not compatible with the notion of strong environmental determination of ability, nor with the irrelevance of IQ to social status.

This table is somewhat daunting, because of its size; Burt (1959), has published a rather smaller one in which he has boiled down social classes into 3 main ones—middle, skilled working class, and the rest. Table 7 shows his main results. It will be seen that when we just

Fathers' status	*Sons' Adult Status* (as percentage of fathers' grouping)			
	I %	II %	III %	Total %
I	51·7	34·5	13·8	100·0
II	23·3	46·9	29·8	100·0
III	13·7	36·9	49·4	100·0

Table 7   *Extent to which sons have same status as fathers*

concentrate on three main social gradings, less than 50% of sons have the same socio-economic status as their fathers. Of sons with fathers in the highest grade, as many as 14% end up in the lowest grade, and a similar 14% of sons end up in the highest grade whose parents were in the lowest. There clearly is a marked migration of children away from the social class into which birth has put them. This phenomenon of social mobility is characteristic of modern societies, and is not confined to England, or to the USA; it is equally characteristic of continental European countries. No understanding of the relation between intelligence and social class is possible which does not take this lack of stability in class into account.

The long-continued studies of Burt have been particularly valuable in throwing light on the relation between IQ and social class. I shall draw rather heavily on his work—partly because few others have given much attention to this problem, partly of course because of the outstanding quality of the design and the statistical treatment in his studies. Burt (1961) begins by making clear two points. During the period of his investigations, which spans some fifty years, there has been no great change in the average level of general intelligence and 'the amount of individual variation about the average level of intelligence has apparently remained fairly constant; certainly it has not declined'. He then goes on to make a third point which will be seen to be of critical importance to our discussion: 'There are appreciable differences in the average level of intelligence in the different socio-economic classes, and in spite of the remarkable improvements in material and cultural conditions, the differences have altered hardly at all during the period in question.'

These are facts on which little argument is possible; it is in the interpretation of the facts that differences of opinion arise. Thus Burt quotes J. Floud and A. H. Halsey (1956) as maintaining the view that the apparent differences between social classes are not in any way due to genetic influences, and that we are instead dealing with 'a hypothesis of near-randomness in the social distribution of innate intelligence'. This implies that the means for all classes would be approximately the same. This view is widespread, not only in this country, but also in the USA where it may be said to be psychological and sociological orthodoxy. Halsey and Floud do not deny that genetic determinants of intelligence exist; in this they differ from other critics whom Burt quotes; their point is the more restricted one that although genetic influences are present, they do not account for the observed differences between social classes. This view has appealed to many people on political grounds, but as we shall see it is quite untenable.

Burt quotes the actual distribution of IQs found in his population, for fathers in the six occupational groups he uses, and for children of fathers in these groups (Tables 8 and 9). The figures given under the heading 'Total' merely represent proportions; thus where the number of fathers in Class 1 is given as three, the actual number examined was nearer a hundred and twenty. The data show in much greater detail the facts of regression set out in Figure 5; in these tables we are able to see also the lower range of ability found in particular classes, as

	50–60	60–70	70–80	80–90	90–100	100–110	110–120	120–130	130–140	140+	Total	Mean IQ
1 Higher Professional									2	1	3	139·7
2 Lower Professional			2	1	8	16	2	13	15	1	31	130·6
3 Clerical				1		16	56	38	3		122	115·9
4 Skilled			2	11	51	101	78	14	1		258	108·2
5 Semiskilled		5	15	31	135	120	17	2			325	97·8
6 Unskilled	1	18	52	117	53	11	9				261	84·9
Total	1	23	69	160	247	248	162	67	21	2	1,000	100·0

Table 8 *Distribution of intelligence according to occupational class: adults*

	50–60	60–70	70–80	80–90	90–100	100–110	110–120	120–130	130–140	140+	Total	Mean IQ
1 Higher Professional						1		1	1		3	120·8
2 Lower Professional				1	2	6	12	8	2		31	114·7
3 Clerical			3	8	21	31	35	18	6		122	107·8
4 Skilled		1	12	33	53	70	59	22	7	1	258	104·6
5 Semiskilled	1	6	23	55	99	85	38	13	5		325	98·9
6 Unskilled	1	15	32	62	75	54	16	6			261	92·6
Total	2	22	70	159	250	247	160	68	21	1	1,000	100·0

Table 9 *Distribution of intelligence according to occupational class: children*

compared with the whole population. Thus none of the fathers in group 1 have IQs lower than 130; none of the fathers in group 2 have IQs below 110. Burt points out that there is considerable overlap: 'in the lowest class of all—that of unskilled workers—some of the brightest members actually display greater intelligence than the dullest members of class 2, the "lower professional".' The correlation between social class and intelligence is therefore far from perfect; it works out at .74, but this figure is of course based on distributions far from normal, and hence its true numerical value is probably lower; the best estimate is more like .5.

For the children, of course, the differences between the class-means are much smaller, as we had already seen in Figure 5. The amount of regression thus found, Burt points out, 'is very close to the value we should expect on the assumption that the correlation between fathers and sons was due chiefly to multifactorial inheritance with assortative mating and incomplete dominance'—that is, just the conditions which on the basis of other data we had actually found to obtain. But if the mean intelligence of the children belonging to each social class deviates less from the mean of the whole population than does that of their fathers, and if the intelligence of the children in a given class varies over a far wider range than does that of their fathers, then these cumulative changes would lead to the complete absence of differences between class-means within five generations unless these effects were counteracted in some way. Such counteraction is of course present in the shape of social mobility; if it were not for the presence of this feature in our society, regression effects would make it mathematically impossible for the means and variances of intelligence in the different social classes to remain constant over the generations.

It is possible to calculate the minimum amount of social mobility which would be needed to maintain a 'steady state', that is to offset the effects of regression.

Burt (1961) calculates this as amounting to 22%, which, considering the necessarily somewhat rough-and-ready nature of such calculations, is not far removed from reasonable assessments of the true value.

This 'true' value, give or take a few points, is around 30%; it cannot be precisely estimated because it is probably never quite steady from year to year, and also because the value depends crucially on the number of subdivisions into 'classes' one employs—the smaller the breakdown, the larger the degree of social mobility.

Taking Burt's breakdown, we can also calculate what would be the ideal amount of social mobility, assuming for the moment that social class were determined entirely by intelligence in some 'ideal' state—a supposition which of course Burt himself disavows; there are other factors, such as personality and motivation, which must play a crucial role in determining a person's social standing. However, taking the six main classes Burt uses in Tables 8 and 9, and regrouping them into three by forming a single group of the upper three middle-class occupations, another single group by putting together the next two working-class groups, and leaving the lowest class as it is, we find that 'only 55% of the population could be regarded as correctly placed if intelligence were the only criterion: nearly 23% are in a class too high, and, with a perfect scheme of vocational selection, ought to be moved down: 22% are in a class too low, and would have to be moved up.'

Thus if intelligence were the only relevant factor in determining social class, social mobility would have to be well above the value of 30% or thereabouts which is actually observed, and which is quite sufficient to maintain a 'steady state' of preserving the IQ difference between classes. In terms of social justice, the failure of social mobility to reach the higher value may be due to two quite distinct causes, one of which is perhaps in the best interests of creating a 'meritocracy', while the other is not. The first cause has already been alluded to; it is simply that qualities of personality, such as persistence, courage, hard work, etcetera, and motivational factors, such as high level of aspiration, in part determine a man's level of achievement, and consequently quite rightly play a part in addition to intelligence in making him a member of the 'meritocracy'. There are also irrelevant factors which work against the creation of a perfect meritocracy; among them are luck, parental influence, and other influences whose main function is to preserve the *status quo*.

But so far we have had no evidence that the rises into, and falls from, higher to lower social class are in fact correlated with intelligence; the figures suggest that something of this kind must have taken place, but that is not the same thing as direct evidence from follow-up studies of individual children. Burt has provided such evidence: he found that of the children with an intelligence below the minimum required for the occupational class into which they were born none rose above it, and more than a third dropped to a lower class. Of those who had an intelligence above the maximum required for their original occupational class over 40% rose to a higher class. He also found motivation a very important factor when coupled with intelligence: 'a good home background . . . was less effective in securing a rise than either high intelligence or strong motivation. Nor was a bad home background so fatal as seems to be commonly assumed.' Nearly a quarter (24%) of those who suffered from unfavourable home circumstances in childhood nevertheless succeeded in rising out of their original class. It is possible to assess the respective force of these factors in determining social mobility, keeping the other factors constant statistically in each case; the correlation between social mobility and intelligence works out at .38, social mobility and motivation at .29, social mobility and home conditions at .17, and social mobility and educational record at .05. It is clear that intelligence is much the most important, followed by motivation;

home conditions and educational record are much less important. These results are not easily encompassed by any environmentalist theory which regards intelligence merely as the consequence of home upbringing and education; they are in perfect agreement with a genetic model which attributes far more importance to heredity than to environment in the causation of intellectual differences.

Burt's study is of course not the only one to come to this conclusion; every research worker who has published data in this field has come to similar conclusions. Let us quote just one further witness, J. H. Waller's follow-up study (1971) of 131 fathers and their 173 sons, studied as a representative sample of white males in the state of Minnesota. 'The data of this study show a consistent association between father-son difference in IQ score and father-son difference in social achievement.' Figure 9 shows the results; graphed in this table is the percentage of sons moving up or down from their fathers' social class according to the amount and direction of difference in IQ between father and son. On the right of the graph are shown the sons who were much brighter than their fathers; almost 80% of these moved up in the social scale. On the left of the graph are shown the sons who were much duller than their fathers; over 80% of these moved down in the social scale. Smaller differences between fathers and sons produced smaller percentages of sons moving up or down in the social scale; note however that the movement is always congruent with the direction of the IQ difference between parents— duller sons sink, brighter sons rise, on the whole. Waller concludes:

> The substantial correlation between the father-son difference in IQ score and father-son difference in social position ($r = +0.368$), and the relationship between the magnitude (and direction) of IQ score difference and the distance (and direction) of social

mobility both support the view that differences in ability provide a 'springboard' that enables individuals to be socially mobile and that to some degree prevents social classes in an open society from congealing into castes.

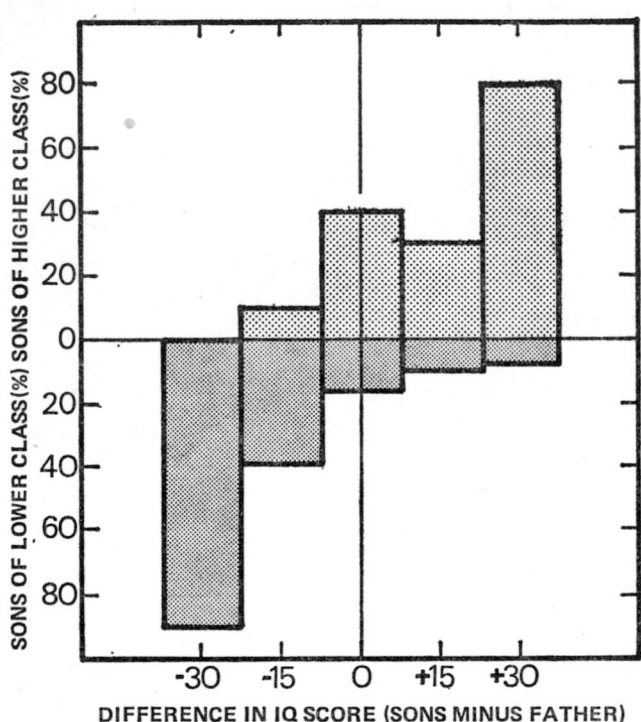

Figure 9 *Percentage of sons moving up or down from their fathers' social class by differences in IQ score*

We must conclude from this discussion that one frequently voiced criticism of IQ tests is quite unjustified; poor showing of children at school, which is correlated with both low IQ and with social class, is due far more to the former than to the latter, and low IQ cannot be blamed on the child's social class, as such. It is the child's IQ, largely inherited, which will in due course determine his social class (in part at least), and which at present determines his degree of success at school (in part at least). Many American studies have shown that schools make very little contribution to the relative standing of children in their various courses; good scholars remain at the top, poor scholars at the bottom, and furthermore the relative gap between them neither increases nor decreases.

What children take out of their schools is proportional to what they bring into the school in terms of IQ. This general conclusion has been fiercely debated, but the facts are firmly against any strong environmental influence being exerted by schools, either in the direction of increasing or decreasing differentials in pupil performance. Many attempts have been made, particularly in the USA, to 'pull up' the performance of children in various 'deprived groups' by various means, or to improve their IQs: operation 'Headstart' has acquired some fame in this connection, but there have been over one thousand such attempts recorded in the literature, almost all of them failures. (By failure is meant in this context that when the progress of the experimental children is compared with that of a control group of children not exposed to the 'enriched' teaching environment, no statistically significant differences appear.) It would be boring to go through accounts of all these studies; it may be more useful to look at just one of them, namely the latest, biggest, and by all accounts the best controlled. The device employed in this study was the so-called practice of 'performance contracting'.

Performance contracting is in some ways peculiarly American; it is based on the principle of piece work— the contractor is paid for the improvement in the school performance of the backward children he contracts to teach, in proportion to the superiority they show over a control group not so taught. According to the Office of Economic Opportunity, which initiated these statistics, performance contracting works as follows: '(1) A contractor signs an agreement to improve students' performance in certain basic skills by set amounts. (2) The contractor is paid according to his success in bringing students' performance up to those prespecified levels. If he succeeds, he makes a profit. If he fails, he doesn't get paid. (3) Within guidelines established by the school board, the contractor is free to use whatever instructional techniques, incentive systems, and audiovisual aids he feels can be most effective. He thus is allowed more flexibility than is usually offered a building principal or a classroom teacher.' Some of the anticipated gains were 'better overall performance—drop-out prevention —integration—individualization of instruction'.

A contractor is a sort of consortium of educational experts, sociologists, psychologists and other 'experts' who have been active in the field and believe on the basis of their past experience that they are able to achieve these aims; also involved are commercial agencies concerned with audio-visual machinery, with automated teaching devices, etcetera. An optimistic atmosphere prevailed in view of the alleged very favourable outcome of the so-called Texarkana experiment, one of the first of the performance contracting studies; this enjoyed such a remarkable success that many people believed that PC might well be the solution to educational inequality. OEO states that 'staff visited Texarkana and saw great promise in performance contracting as a means to help poor children achieve the same results from classroom effort now achieved by non-poor students.'

Critics have had a field day with Texarkana, drawing attention to several pitfalls into which the contractors and OEO had fallen. These included improper selection of groups, 'teaching to the test', and failure to take regression into account. Texarkana seems to have fallen into them all and OEO were keen that no criticism should attend the large-scale testing of the claims of performance contracting they were contemplating. Let it be stated at once that under the benevolent eye of the Battelle Memorial Institute, which supervised the experiment, no major infringement of the proper design seems to have taken place, and that in spite of minor deviations from the original plan the results constitute the biggest and best test of educational claims which has ever been carried out.

Thirty-one companies bid for the contracts: six unlucky ones won the bid. Professor Page (1972), who has reviewed the programme, points out that it was a punishing experience for all six, and three or four are now out of business. The six contractors reflected the most persuasive and varied practices in applied psychology.

Each hired company would apply its principles in three different school systems (18 in all), widely separated throughout the US, across six grade levels . . . and across every major low-achieving ethnic group, from poor urban Anglos to Alaskan Eskimos. There was to be, furthermore, generalization across the two most important educational skills: reading and arithmetic. Over 25,000 students would participate, and over six million dollars would be spent. . . . The 18 participating districts were carefully chosen, and the subjects, both experimental and control, were from the worst-performing schools in the districts. And from each school, for each grade, the 100 students furthest behind their grade levels would serve as subjects. The resulting composite of subjects is very representative of the target population.

What did the contractors actually do? The answer is difficult to give in detail, as the different contractors used different methods, or combinations of methods; furthermore, only preliminary write-ups of the experiment are available, and these do not go into much detail. Common to nearly all the contractors was a belief in the importance of motivation. Lack of motivation, so common in the poorer children, was to be overcome by incentives —'e.g., trading stamps, that they would use for gifts, and free time during the class period that they could use to read magazines, listen to records, or other recreational activities.' Audio-visual aids played a great part in the 'new' teaching, both to increase motivation, and to put over the material more interestingly. Individual attention to children was another aim, achieved by having more teachers and helpers, and by 'programmed teaching machines'. Add to all this the inevitable 'Hawthorne effect', that is the usual finding that concerted attention to a group by devoted social scientists usually produces by itself a marked improvement in their performance (even when the changes made are normally detrimental to performance!) and the ground would seem well prepared for a tremendous difference in performance between the experimental and the control children. On the basis of all that modern educationalists (particularly American ones) believe in, the contractors would seem to have been on an easy wicket. Yet they were all out for a duck; there were no differences between experimental and control children at the end of the experiment.

Let us look again at the actual contract they signed. Payment was to be made only for a student who gained one school year of growth between the fall pre-test and the spring post-test. As Page notes:

> the design eliminated ordinary upward regression, teaching of the test, and biasing of evaluation—all three of them the bane of sound inference, but the comfort of those determined to report favourable results. What was left was essentially improvement in the actual, general, *transferable* skills of reading and arithmetic.

In expecting to be able to make these under-achieving youngsters cover a year's advance in six months the contractors would seem to have been somewhat optimistic; with the usual type of instruction, these boys and girls would have been expected to advance by about .7 of a year's growth in a whole year—experience teaches that those who are behind already fall more and more behind in the course of time. To reverse this would really be to work a miracle; the miracle did not happen.

How did it come about that contractors, teachers, and OEO walked into the trap so happily, without noticing the warning portents? There can only be one answer—they really believed that the non-achieving students were disadvantaged only in their previous experiences; that the ordinary schools were very ineffective teachers; and that the application of modern educational principles would lead to marked improvements in achievement. One cannot deny the contractors the courage of their convictions: they were ready to back up their claims with their own money, and to compete in a race where great care was taken to see that no cheating was possible. Nor can one deny that OEO deserves much credit; they had been severely criticized previously over the design and evaluation of Texarkana and other experiments, and they were determined to put their beliefs in performance contracting on the block. The fact that this led to their heads being chopped off should serve as a lesson to us. (Several heads at OEO were rolling when the results came out; this seems unfair on those responsible for a very important experiment!)

It may be worth while quoting the report prepared by OEO themselves, as any attempt to paraphrase might be regarded with suspicion. Speaking of the experimental and control groups, OEO states that the

difference in gains was remarkably small in all 10 of the grade/subject combinations for which this analysis is appropriate. In half of the 10 cases, there was no difference at all between the gains of the experimental and control groups. In four of the cases there was a difference of as much as two-tenths of a grade level. These overall differences are so slight that we can conclude that performance contracting was no more effective in either reading or maths than the traditional classroom methods of instruction. . . . The performance of students in the experimental group does not appear disappointing just because students in the control group did unexpectedly well. In fact, neither group did well. In only two of the 20 possible cases was the mean gain of either the control or experimental students as much as one grade level. . . . In all cases, the average achievement level of children in the experimental group was well below the norm for their grade and in all cases, in terms of grade equivalents, the average slipped even further behind during this year. . . . There is no evidence that performance contracting had differential results for the lowest or highest achieving students in the sample. . . . Not only did both groups do equally poorly in terms of overall averages, but also these averages were very nearly the same in each grade, in each subject, for the best and worst students in the sample, and with few exceptions, in each site. Indeed, the most interesting aspect of these conclusions is their very consistency. This evidence does not indicate that performance contracting will bring about any great improvement in the educational status of disadvantaged children.

This clearly demonstrates the failure of performance contracting; but it would be wrong to dismiss the results casually as simply the failure of a new-fangled American invention which we would not wish to imitate in any case. The results cut much deeper.

Educationalists, psychologists, sociologists, experts of all kinds have over the past twenty or thirty years been suggesting new and allegedly 'better' methods of teaching; these new methods, it is suggested, would achieve what traditional methods have failed to achieve. Now the OEO experiment has given these new methods a chance to show what they can in fact do; the outcome has been disastrous. It must be seen in the light of many reports published in the same twenty or thirty years which apparently showed triumphant successes for this method or the other; practically all these experiments were badly designed and permitted regression effects, teaching the tests, and other faults to invalidate the conclusions. Now we know that even in combination these methods do not advance us in any way, shape or form; we are truly 'back to the drawing-board'. Let us hope that the lesson will be well and truly learned; let us never again be persuaded of the excellence of this or that 'new' method without proper, well-controlled, adequately designed experiments which contain a suitably selected control group. Education is second only to psychoanalysis in making claims for untested and untried methods, and in refusing to put these methods to a proper test. Perhaps one effect of the OEO experiment will be a greater resistance to claims made on the basis of theoretical assumptions, instead of proper experimental demonstration.

The conclusions of the OEO report agree well with those of the Coleman report (1966), an enormous and very careful survey of the American school system. Very little of the variance among schools in scholastic achievement was due to differences in school facilities, including variables such as physical facilities, class size, curricula, teacher salaries, experience and qualifications, special services, etcetera.

> Differences in school facilities and curriculum which are the major variables by which attempts are made to improve schools, are so little related to differences in achievement levels of students that, with few exceptions, their efforts fail to appear even in a survey of this magnitude.

Does modern psychology really have nothing to teach us in this field? Must we really go back to Victorian times for our teaching methods? Let me put forward one view which I know is not shared by many educationa-lists, but for which there is now quite a lot of evidence. According to this view it is quite idle to look for any method of teaching which would be equally suitable for different types of personality; wisdom would lie in group-ing children according to their personality, at least as much as streaming them according to their abilities. Thus, to take but one example, there is evidence that intro-verted children benefit particularly from the use of 'teaching machines'; programmed instruction of this sort puts off the extroverted child who is more person-oriented, and requires personalized instruction. The new 'discovery' methods of teaching go down well with the extroverted child, but put off the introverted one. Thus it is possible to introduce into a class a combination of machine teaching and discovery method which leaves the class no better off and no worse, taken as a whole; had the discovery methods been used with the extro-verted children, the machine teaching methods with the introverted, there might have been a marked improve-ment in their performances.

What is at fault, in general school practice as well as in the work of the performance contractors of the OEO experiment, is the tacit assumption that all children are alike, and that what is good for one is good for all the others. No good teacher believes this for a minute, of course, and every good teacher attempts to gear his methods to the individual child; nevertheless, this is made tremendously difficult by failure to concentrate research on the precise differences which are relevant to school learning, and by failure to group children accord-ing to relevant personality type. One can see the diffi-culty of performance contractors in coming to grips with this problem, but this hardly excuses their failure even to realize its existence.

This brings me to the last point in this melancholy story. The contractors, as well as OEO, had become victims of their own propaganda, according to which environment is all-important, and heredity counts for nothing. As so many different experiments have shown, heredity is twice as important as environment in determining IQ, and IQ, whether we like it or not, is still the most important single cause for success at school. This should not lead us to abandon the search for better methods of education, particularly for the duller child; the stress on the need for compensatory education is a moral issue, which cannot be sidetracked by any factual argument. But it should warn us that the path will not be easy, and that we have hardly made much progress along it at the moment. If we paid more attention to the facts of genetic influences, we might advance a little faster.

The myth that education can offset differences in innate ability dies hard; most people seem to have strong defence mechanisms which make them immune to the facts. It may be useful to take one specific example of such an educational myth, and look at the facts regarding it. The myth: that smaller classes help children learn more. This seems so much part of enlightened common sense that few people, parents or educationalists, seem inclined to look for evidence. We are constantly urged by teachers' associations to make smaller classes a priority, on the grounds that doing this would help the children to reach higher standards. Few people ever ask themselves whether this rosy picture is in fact at all related to reality. The answer, alas, seems to be in the negative; smaller classes seem to have exactly the opposite effect, that is children in smaller classes do not learn as well as children in larger classes.

Joyce Morris had originally found that children in the larger classes of infants and juniors had better reading attainments than children in smaller classes. Stephen Wiseman considered a sample of 25% of Manchester Primary schools, constituting a proper sample; he too found a small but positive correlation between size of class and educational attainment. The National Child

Development Survey, which followed up a complete 'cohort' of children born in one week in 1958, found that even taking account of school size, length of schooling, parental interest and occupation, the children in larger classes still did better than those in smaller classes. These were mainly young children of primary school age; a study by Marklund in Sweden, employing a national sample of thirteen-year-olds, found very similar results. Grouping class sizes into intervals—16 to 20; 21 to 25; 26 to 30, and 31 to 35, he found that attainment was highest in the classes of 26 to 30 and lowest in the classes of 21 to 25. When groups were combined, the classes with 26 to 35 pupils in them were better than those with 16 to 25 pupils. Similar results come from recent studies by the Inner London Educational Authority; again the children in bigger classes did better. There are several other studies, but these may suffice to make the reader wonder why educational authorities continue to work for smaller classes; are they in fact impervious to facts? Are they trying to lower the performance and attainment of their pupils? If more money is available for education, the provision of smaller classes surely is not one of the more obvious ways in which one would be inclined to spend it; yet the power of the myth is such that that is probably exactly what educational authorities will try to do.

Size of class has been chosen as one of many possible examples to illustrate the strong beliefs which many educationalists have in environmental manipulation of variables in an effort to influence the educational attainments of children, particularly of 'deprived' children, for the better; there is usually very little hard evidence to back up these beliefs. When evidence is available, it is most frequently in the direction of 'no difference', that is it suggests that the variables alleged to produce improvement have in fact no effect one way or the other. Occasionally, an effect is demonstrable, but this is as likely to be in the wrong direction as in the predicted direction—as in the case of size of class. Myths are particularly likely to grow in the educational field because so many of our egalitarian hopes and aspirations are tied up with education; education seems to be the great equalizer, above all others.

Yet the facts are different: we are not likely to produce any greater degree of equalization than that already reached by any traditional modification of the schooling system. It is interesting that this conclusion is now more widely accepted than it used to be, although even now this acceptance is very partial. In evidence of the fact that a wholly environmental approach is still very much educational orthodoxy, consider the publications of the National Survey of Health and Development, and more particularly those of the National Child Development Study. Davie, Butler and Goldstein have just published the second report in this follow-up study of the 1958 cohort of children born in one week of 1958, and a look at its contents is very instructive in this respect.

This book looks in the minutest detail into the causes of the educational performance of these children, and the causes of any differences that appear. 'One of the most striking features which emerges from the results we have presented is the very marked differences between children from different circumstances, which are already apparent at the age of seven. . . . Furthermore, the most potent factors were seen to be located in the home environment.' Note that there is no mention in this (or anywhere in the book) of genetic factors, or even of intelligence at all; 'intelligence' and 'heredity' are both missing from the index! Here truly we have Hamlet without the Prince of Denmark; a book of 586 pages containing over 300 tables, devoted to the discovery of the main causes of educational differences between children, which does not even mention what we have seen are the main causes of such differences! It may be invidious to have taken this particular example when so many others could have been chosen instead; the tragedy is that this study is typical of current research in the educational field.

In dealing with the effectiveness of environmental variables, psychologists have changed their stance in recent years, and have abandoned the older notion that education in the classical sense can do very much to alter differences in IQ between children coming from different backgrounds. It is more fashionable now to stress either antenatal influences, or environmental factors occurring during the first few years of life, that is during the pre-school period. Antenatal influences may perhaps be ruled out as a factor of wide importance, omitting the occasional case of toxic poisoning, brain damage, asphyxia, etcetera. We have already considered the 'critical period' theory of malnutrition, according to which the effects of malnutrition should be particularly marked if it occurred during antenatal periods or shortly after birth; the results of the Dutch famine tests have falsified this hypothesis. The foetus is extremely well protected in the mother's womb, and there is no evidence that events likely to occur during the average mother's child-bearing life produce any marked influence on the child's IQ. Joffe (1969) has reviewed the whole literature, both experimental and observational, on prenatal determinants of behaviour; he finds much evidence of such influence as far as emotional behaviour is concerned, but not as far as intelligence is concerned. While the question is not completely closed, the best available evidence does not suggest that there is much of importance to be found here.

The position is quite different with respect to environmental influences which occur during the first few years of life. Bloom (1965) has given a valuable summary of the evidence, starting with the view that 'the long-term effects of extreme environments may affect the IQ to the extent of . . . about 20 IQ points. . . . This is about the figure cited by Burks as the effect of extreme environments.' (It will be remembered that we have already encountered Burks' important study of the correlation between the IQ of foster children and various environmental determinants of intelligence.) Bloom also estimates the extent of development of intelligence at various ages from many figures given in the literature: according to this estimate, 50% of total adult IQ is already developed by the age of 4, another 30% accrues from 4 to 8, and the remaining 20% is consolidated by the time the adolescent reaches the age of 17. These figures are far from accurate, of course, but may serve as a provisional guide to the approximate differences in intellectual development which occur from early babyhood to adolescence. Bloom presents a table which is of much interest, even though the actual figures should not be taken too seriously (Table 10).

Age	Per cent of adult IQ	Deviation from normal in IQ points			Total difference
		Deprived	Normal	Favoured	
Birth to 4	50%	−5	0	+5	10
4–8	30%	−3	0	+3	6
9–17	20%	−2	0	+2	4
Total	100%	−10	0	+10	20

Table 10 *Differences between environmentally deprived and favoured children in IQ, showing strength of effect of environment at different ages*

This shows that if we compare the intellectual development of the most deprived and the most favoured children (as far as environment is concerned) the total difference of 20 points may be subdivided into three portions: 10 points are produced by environmental differences occurring before the child is 4 years old, 6 points are produced during the period from 4 to 8 years of age, and the remaining 4 points accrue during the next 9 years. If this is even approximately a correct picture, then we can see why education in the sense in which it is usually talked about (that is secondary education, and perhaps the last two years of primary education) has so little effect; it cannot produce more than 4 points of IQ difference between the most deprived and the most favoured! Bloom cites in detail the evidence on which his conclusions are based; it would take us too far afield to follow him in this. But we may note the major environmental influences which he singles out from the empirical literature as influencing IQ.

These are divided into three main groups, the first of which is called 'Press for Achievement Motivation': this is concerned with the nature of intellectual expectations of the child, the nature of intellectual aspirations for the child, and the nature of rewards for intellectual development. The second main group deals with 'Press for Language Development': it deals with the emphasis placed on the use of language in a variety of situations, the opportunities provided for enlarging the vocabulary, and emphasis on correctness of usage, as well as the quality of language models available. The third and last group of variables is entitled 'Provision for General Learning'; this refers to opportunities provided for learning in the home, opportunities provided for learning outside the home, availability of learning supplies; and the nature and amount of assistance provided to facilitate learning in a variety of situations.

There are two main criticisms which one may make of such a list, and of the research on which it is based. In the first place, many of these variables are simply reflections of the IQ of the parents, and cannot therefore be regarded as independent of the genetic determination of the child's IQ. Bright parents have children who inherit the predisposition to high IQ, and they also provide opportunities for learning, etcetera. These factors require careful sorting out, and they have not received this in the literature quoted by Bloom. In the second place, these variables would refer to what geneticists call between-families variance; it is doubtful if on these variables children within a given family would differ very much. But we have already quoted decisive evidence to show that between-families variance in intelligence is small or non-existent; neither is there much evidence for interaction effects between heredity and environment. This reinforces the first point made; it seems unlikely that the variables mentioned in the list can be regarded as separate environmental factors determining the child's IQ independently of the parents' IQ. While it is probably true that environmental factors are most important in affecting IQ during the child's most formative years, the actual factors which are active in this situation still remain to be sorted out. This is a difficult and complex task which cannot be undertaken without at the same time measuring hereditary factors; the two sides of the coin are truly inseparable.

We have seen in this chapter how closely educational attainment depends on IQ, rather than on social class; social class is indeed correlated with educational success, but mainly through the fact that social class itself is determined quite strongly by IQ. We will consider in a later chapter whether better methods cannot be devised to make use of the leeway given us by the 20% of the total variance which is contributed by environment; it has been made clear that at present we are largely groping in the dark as to what we can do to improve the performance of 'deprived' children. We will then also consider whether anything can be done to affect the genetic component of IQ; it would be quite incorrect to imagine that we must accept this as given.

But before closing this chapter we must discuss in some detail one further point which is often brought up in connection with the discussion of social class, intelligence, and educational achievement: the so-called Pygmalion effect, or the importance of 'self-fulfilling prophecies' in the educational field. This point was brought to public attention by the publication, in 1968, of a book entitled *Pygmalion in the Classroom: Teacher Expectation and Pupil's Intellectual Development,* by Robert Rosenthal and Lenore Jacobson. The experiments reported in this book purported to show that a pupil's IQ could be made to rise if teachers were informed (incorrectly) that that particular pupil was a 'spurter', that is that he had a high IQ. The results of the experiment, in which all the pupils were tested before, during and after two years' teaching (four testings in all), were interpreted as showing '. . . that teachers' favourable expectations can be responsible for gains in their pupils' IQs and, for the lower grades, that these gains can be quite dramatic'. These 'findings' were widely quoted by the press, and far-reaching conclusions drawn; thus one report read as follows: 'Here may lie the explanation of the effects of socio-economic status on schooling. Teachers of a higher socio-economic status expect pupils of a lower socio-economic status to fail.' Another said that 'the findings raise some fundamental questions about teacher training. They also cast doubt on the wisdom of assigning children to classes according to presumed ability, which may only mire the lowest groups into self-confining ruts.'

The book has been reviewed by several outstanding experts in the field, and their unanimous conclusion has been that the data presented have in fact very little evidential value; there may indeed be such a thing as the 'Pygmalion effect' (the very name is a curious misuse of a classical Greek legend) but the results of the experiment do not provide any evidence in its favour. Professor R. E. Snow (1969), who has published what is perhaps the clearest and most incisive review of the book, concludes it by saying of the Rosenthal and Jacobson book that 'their study has not come close to providing adequate demonstration of the phenomenon or understanding of its process. *Pygmalion,* inadequately and prematurely reported in book and magazine form, has performed a disservice to teachers and schools, to users and developers of mental tests, and perhaps worst of all, to parents and children whose newly gained expectations may not prove quite so self-fulfilling.'

What, more particularly, is wrong with the study? When a comparison is made between first and last testing, there is practically zero difference between experimental and control groups in mean gain for all classes except the two youngest, where the experimental gain apparently did exceed the control gain. This is interesting because the test used for the comparison does not in fact provide adequate norms for the younger children, particularly those coming from poorer homes. The test does not provide norms below an IQ level of 60, but for Grade One IQ means are quoted as 47 and 31 respectively for 16 middle and 19 low track control subjects, and 54 and 53 for 4 and 2 experimental subjects. The average for all first grade children is 58. Snow raises the interesting question: 'Were all these children actually functioning at imbecile and low moron level? More likely, the test was not functioning at this age level. . . . To obtain IQ scores as low as these, given reasonably distributed ages, raw scores would have to represent random or systematically incorrect responding.' One child had successive verbal IQs of 183, 166, 221, and 168, although the test used does not have norms above 160! Thus we see that an inappropriate test was used, which gave nonsensical results precisely for those age groups where positive results are claimed, and for which the test was most inappropriate. For those ages where the test was appropriate, no differences were in fact found between experimental and control children. There are innumerable statistical errors and malpractices, which are faithfully pointed out by Snow, but which will not be repeated here. There are also malpractices of reporting:

> score distributions are not given. Graphs and tables are frequently misleading: some show only differences between different scores, where basic data are

6

> not available in the book . . .; some fail to indi-
> cate the small sample sizes on which impressive
> percentages are based . . .; some use microscopic
> scales to over-emphasize practically insignificant
> differences . . .; and some display floating zero
> points and elastic scales, making comparisons from
> one scale to another difficult.

We need only add that several attempts to replicate the
findings, using proper designs and orthodox statistics,
have completely failed to find any evidence for this
alleged 'Pygmalion effect' to make it clear that the
evidence for its existence rests on a very insecure foundation indeed.

What is so interesting is the fact that this inadequate
experiment became famous overnight; its conclusions
have been widely publicized not only in the USA but
also in the UK and in many other countries; and the
results have been widely welcomed as providing scientific
evidence with which to clobber psychologists and geneticists who dared to avow the importance of genetic determinants of intelligence.

It would seem that myth begets myth; the myth that
education can appreciably decrease the gap in intelligence between members of the 'deprived' classes and
members of the 'advantaged' classes, when shown to be
without much foundation, seeks salvage by breeding
other myths, such as that of the 'self-fulfilling prophecy'
that it is *because* children are labelled 'dull' or 'bright' by
IQ tests that they perform poorly or well at school. It is
curious that this alleged bias on the part of the teachers
goes counter to the bias which most of them would
probably acknowledge, that is that of trying particularly
hard to bring the dull children up to the level of the
bright; many teachers would seem to regard this other
bias as almost an ethical imperative. However that
might be, there clearly is no worthwhile evidence to
suggest that the data reviewed in this chapter are in any

sense to be reinterpreted in terms of the 'Pygmalion' effect; to date there is no evidence for the existence of such an effect, and plenty of evidence to support the null hypothesis, that is, the view that knowledge of the pupil's IQ on the teacher's part does not affect the subsequent IQ measures of the pupil. If indeed bright children are treated by the teachers in such a way as to increase their IQs, while dull children are treated by the teachers in such a way as to decrease their IQs, then these two groups should pull further and further apart; the evidence shows that they do not (Jensen, 1973).

CHAPTER 5

Personality, mental illness and crime

As we have noted before, the distinction between the cognitive or intellectual side of human nature, and the emotional and conative side, had already been commented upon by Plato and other Greek writers. The question arises of whether heredity plays an equally vital part in causing differences in these emotional aspects of personality as it does in the intellectual aspect. Important as intelligence is in problem solving and in adjusting one's behaviour to prevailing social conditions, the emotional side of human nature exerts a tremendously important effect on our behaviour which was not fully recognized by philosophically-minded psychologists who felt more at home with the discussion of cognitive issues; it was the impact of mental illness and criminal behaviour on the public consciousness that first drew attention to the importance of non-cognitive factors in human conduct. If crime does not pay, then why do so many people indulge in criminal activity? If mental illness were an isolated phenomenon, then one might disregard the challenge it presents to any purely intellectual interpretation of human behaviour; however, mental illness is so widespread, as we shall see, that there will be few readers of this book in whose family there is not at least one member who has been in treatment for a psychiatric disorder. The very concept of mental illness, almost by definition, implies conduct which goes counter to cognitive rules recognized by the majority; a person who believes himself to be Jesus, or Napoleon, in spite of the strong evidence which contradicts claims of this kind, is clearly not acting rationally, and it is impossible to explain his conduct in cognitive terms.

In this chapter we shall first look at the role which heredity plays in the causation of criminal and mentally abnormal behaviour; after that we shall turn to the role which heredity plays in causing variations in more normal personality patterns. To introduce the methodology used in studies of socially recognized patterns of misconduct we shall take criminality first; the criteria used are simpler than those used when we are dealing with mental illness, and there are many complications in dealing with the latter which do not arise with crime. The main advantage of using criminality is the clear-cut nature of the criterion; the laws of the country define crime fairly unambiguously, and commitment to prison is an unambiguous index of criminality. At least, so it would appear at first; we will soon have to note complications arising even in this apparently simple paradigm.

The method used to study the strength of hereditary determination of criminal conduct derives from the underlying dichotomy of identical and fraternal twins, and is similar to one of the methods already discussed in our dealing with the inheritance of intelligence. There we compared the correlation between identical twins for IQ, and compared it with that obtained from fraternal twins; the fact that the correlation is higher for identicals than for fraternals can be used as evidence for the influence of heredity. IQ of course is a continuous variable which can be fairly accurately measured; criminality is more an all-or-nothing phenomenon, that is a person is or is not a criminal in the sense of having been at least once sentenced to prison because he had committed a crime as defined by the law of his country. It might be said that this is not quite accurate, and that criminality is surely a graded phenomenon, too; a person who has committed one crime is less 'criminal' than another who has committed several dozen crimes. This is perhaps true in general, but suppose the one crime is rape followed by callous murder of the woman involved, while the dozens of crimes are merely burglaries committed over the years? It is the difficulty involved in assessing the seriousness of crimes, and the inevitable subjectivity involved in assigning numerical values to the 'criminality' of people incarcerated for different numbers of qualitatively different crimes, which has led investigators to prefer the all-or-none rule, which involves no subjectivity on the experimenter's part; he merely lays down his criteria, and rigidly follows them in his investigation. This may make it more difficult to find evidence for hereditary determination of criminal conduct, but it guarantees that if such evidence is found it is not artificial and due to his own subjective judgments. Given then that most investigators have used this all-or-none criterion (about which we shall have more to say later on) the method of correlation becomes impossible, and we are left with what is known as the method of concordance.

Typically, what the investigator does is to enter a given prison and seek out (either in person or by way of the register) all those prisoners who are twins; these prisoners are the probands in the investigation. The investigator then seeks out all like-sexed co-twins of the probands and ascertains two things: are the twins identical or fraternal, and is the co-twin concordant (is he, too, in prison, or has he been in prison at any time) or discordant (that is has he never been in trouble with the law). If heredity plays an important part in causing delinquency, then we would expect identical twins to be concordant more often than fraternal twins. What are the facts?

J. Lange (1929), a German psychiatrist, was the first to investigate this matter. His book *Crime as Destiny* (with a commendatory foreword by J. B. S. Haldane) showed that out of thirteen pairs of identical twins, ten were concordant for criminality, while out of seventeen pairs of fraternal twins, only two were concordant. Table 11 shows his results, as well as those of eight later investigators; this table reports results from studies involving two hundred and thirty-one pairs of identical twins and five hundred and thirty-five pairs of fraternal twins. It can

Author:	Identical:			Fraternal:		
Lange, 1929	10/3	=	77%	2/15	=	12%
Legras, 1932	4/0	=	100%	0/5	=	0%
Rosanoff, 1934	25/12	=	68%	6/54	=	10%
Krantz, 1936	20/11	=	65%	23/20	=	53%
Stumpfl, 1936	11/7	=	61%	7/12	=	37%
Borgström, 1939	3/1	=	75%	2/3	=	40%
Yoshimasu, 1961	17/11	=	61%	2/16	=	11%
Hayashi, 1967	11/4	=	73%	3/2	=	60%
Christiansen, 1968	27/54	=	33%	23/340	=	6%
Total:	128/103	=	55%	68/467	=	13%

Table 11   *Concordance rates for criminality: identical and fraternal twins*

hardly be said that the numbers involved do not enable us to come to any agreed conclusion. It will further be seen that while the percentages vary quite a bit (which is not surprising in view of the fact that some studies used only very few cases), nevertheless in every study the identical twins show greater concordance than the fraternal ones, regardless of the country in which the study was carried out. The overall comparison, using all available data, is given in the last line: 55% of identical twins are concordant, but only 13% of fraternal twins. Thus concordance is found over four times as frequently in identicals as in fraternals, a finding which seems to put beyond any doubt the point that heredity plays an extremely important part in the genesis of criminal behaviour. It is interesting to note in this connection that serious crimes show about twice as high a concordance rate as minor crimes, and that as far as juvenile delinquency is concerned there is not very much difference in the concordance rates for identical and fraternal twins.

Readers may feel that these data are subject to serious errors which may reduce the attention which ought to be paid to the findings. In many of the older studies the diagnosis of zygocity was less than perfect, and this may have allowed a number of twin pairs to have been misplaced. Criminality, if defined in terms of imprisonment, is obviously subject to serious errors of determination; it is well known that only a proportion of all crimes committed are brought to the attention of the police, that of those brought to the attention of the police only a proportion are solved, and that of those guilty of these crimes only a proportion are in fact sent to prison. Thus our data are clearly not very accurate, and this may strongly influence the final results reported. These objections are of course perfectly sound, but we must try and determine the direction in which such errors would bias the final outcome. The answer is simple: the more inaccurate the data, the less will they support the genetic hypothesis! In other words, the final result reported above (that concordance rates are over four times as high for identical as for fraternal twins) is a *minimum* estimate of the true figure; if the various sources of errors could be eliminated, then a much greater difference would be likely to appear as compared with that actually found.

It is easy to see why this must be so. Identical twins are more concordant than fraternals; if we erroneously count a pair of identical twins (who have a high chance of concordance) as fraternal, then we lower the probability of finding concordance among identicals, and increase the probability of finding it among fraternals. The opposite happens when a pair of fraternal twins is misdiagnosed as identical. The twins, having a low chance of concordance, are now counted as members of a group having a high chance of concordance, and will accordingly lower that chance. Thus misdiagnosis (which would under no circumstances be expected to run over

5%) would lessen the true difference between identicals and fraternals, and thus work in a direction opposed to our hypothesis. The same is true of proper criminals being erroneously diagnosed as non-criminals, that is non-concordant. In view of the fact that identical twins are more frequently concordant, those criminals who are wrongly diagnosed as non-criminal will be more likely to reduce the number of identical twins classed as concordant, so working against the theory of criminal heritability. Each such misdiagnosis lowers the degree of differentiation between identicals and fraternals.

This problem of proper diagnosis of criminality is a very serious one, so much in fact that the observed differences between identicals and fraternals appear miraculously high. Consider the following possibilities of error: (1) The co-twin is a criminal but has not been caught. This is known to happen quite frequently, but there is nothing that can be done about it to correct our data. (2) The co-twin has not yet committed a crime, but will do so in the near future. Again, there is no way of taking this fact into account. (3) The co-twin has committed a crime and has been imprisoned, but in another country, so that his record is not available to the investigator. Thus an English twin might have been imprisoned in the USA or Australia, a German twin in Switzerland or Austria, and there is no guarantee that this fact would be known to the police in his own country, particularly if he committed the crime under a false name. There are innumerable possibilities of missing the criminal nature of the co-twin, once the proband has been located, and each such 'miss' is likely to reduce the differential between identicals and fraternals.

There is another criticism of this method which also indicates that the true heritability of criminality is probably much higher than that indicated by the figures. Crime is usually defined in social terms; the word has no absolute connotation. What is a crime in one country may not be one in another, and what is a crime in one country at a given time may not be a crime in the same country at another time. Homosexuality is a good example; this is considered a crime in some countries, but not in others, and it was considered a crime in England a few years ago, but is not now. Such varying definitions must reduce the clarity of the differences observed in concordance rates for identical and fraternal twins, which would presumably be greatest for 'core' crimes such as murder, theft, burglary, assault, etcetera. In actual fact this last point is probably of minimal importance; the great majority of criminals included in the studies cited were in fact imprisoned because of 'core' crimes, and other types of crimes seldom if ever make an appearance in the records.

Of particular interest in the literature is the fact that quite frequently concordance extends to the particular type of crime committed by the two twins—even when they were separated early on in life and were ignorant of the other twin's criminal behaviour. Thus arson in the one twin might be prognostic of arson in the other; assault and violence in the one might go with assault and violence in the other; a propensity to sex crimes in the one might also be found in the other. This specificity is particularly convincing proof of the importance of heredity, and readers are invited to read some of the descriptive studies published by Lange in *Crime as Destiny*, in which he points out these similarities. Something very similar has been found with neurotics; here also concordance is much closer even for specific symptoms than for broader classifications. Some of the similarities of behaviour discovered for concordant identical twins are quite uncanny, and give purpose to the title of his book —*Crime as Destiny*. (This title of course carries implications of complete genetic determination which are unjustified; it would be just as foolish to overlook the importance of environmental factors as it would be to overlook the importance of the genetic contribution.)

It is noteworthy that the observed differences between

identicals and fraternals are almost impossible to explain in terms of the variables which sociologists usually postulate as 'causes of crime', such as poverty, poor education, unemployment, living in marginal districts, and so on; clearly, twins would be likely to be exposed to identical conditions, and this would not be dependent on their being identical or fraternal twins. The data quoted do not prove that the sociological conditions are irrelevant to crime; they merely prove that they interact with powerful inherited predispositions to produce criminal conduct. Crime cannot be understood in terms of heredity alone, but it can also not be understood in terms of environment alone. Thousands of youngsters grow up in conditions which according to theory are most conducive to crime, without succumbing; thousands of other youngsters grow up in conditions which according to theory are unlikely to lead to crime, and yet they adopt a criminal mode of life. Such facts make any purely environmental theory meaningless, just as any purely hereditary theory would be absurd; both heredity and environment play their part, and the data suggest that heredity is at least as important as environment.

Twin studies are not the only support for the genetic-environment interaction hypothesis; adoption studies come to very similar conclusions, and are if anything even more telling. Two important adoption studies have recently been carried out, one in Denmark, and the other in Iowa. In the first of these, Schulsinger (1972) compared 57 psychopathic adoptees with 57 non-psychopathic controls, equated for sex, age, social class, and in many cases neighbourhood of rearing and age of transfer to the adopting family; carefully defined criteria for psychopathic behaviour were used in this study. Next, the investigator examined the case records of the biological and adoptive relatives of the psychopathic and control subjects. In spite of the fact that adoption took place at an early age, there were no differences whatever between the adoptive families of the psychopathic and the control groups; when it came to the biological family members of these groups, however, relatives of the psychopathic boys showed an incidence of psychopathy two and a half times as great, and an incidence of mildly psychopathic behaviour also two and a half times as great, as was found in relatives of the control boys. In other words, the psychopathic boys had taken after their biological parents, not their adoptive parents.

In the Iowa study (Crowe, 1972), interest was not in diagnosed psychopathy, but rather in actual records of arrest. Here the investigator started off by locating 41 female criminal offenders who were inmates of a women's prison reformatory, and who had given up their babies for adoption. At the time of study, they had produced 52 offspring, ranging in age from 15 to 45 years. A properly matched control group of 52 offspring from non-criminal mothers was also studied; these, too, had of course been given up for adoption. It was found that the offspring of the criminal mothers had had more criminal arrests, and had also received a much greater number of convictions; these differences were fully significant statistically. They also had more 'moving traffic violations' recorded against them; for instance, the offspring of the criminal mothers had 19 speeding convictions as compared with 8 for the offspring of non-criminal mothers. Of particular interest is a comparison of the offences committed by the mother and child; there is a clear tendency for these to be similar in nature. In this respect, the study agrees well with Lange's original work with twins, where he also found concordance in the type of crime committed between the identical twins studied. It is impossible at the moment to give a numerical estimate of the relative contribution of heredity and environment to the total variance, but about the importance of heredity there can be no doubt.

This statement is sometimes pilloried by Marxist writers who argue that crime is a feature of capitalist and pre-capitalist society, and that it will disappear in a truly socialist society. I am not competent to argue whether this statement is true or not, particularly as the proper experiment (which presupposes the existence of a proper socialist state) has not yet been carried out. There is no doubt that prisons have not been abolished in Russia, although it can always be argued that Russia is not a truly socialist country. I suspect that the definition of a truly socialist country will give rise to considerable controversy, and that we may end up with a circular definition in which a truly socialist country is defined (partly) as one in which there is no crime. However that may be, the argument is quite irrelevant to the demonstration that in a given country (England, or Germany, or the USA) at a given time, under a particular system of government and production (democratic capitalism), heredity plays an extremely important part, in interaction with environment, in determining who shall commit criminal acts, and who shall not.

We have seen in an earlier chapter that statements about the relative contribution of heredity and environment to the total variance can only be made with respect to a particular population; it is perfectly possible that by changing completely the social and political conditions under which people live, the relative importance of heredity and environment can be altered drastically. One would imagine that by creating greater equality in a society, the importance of heredity would be increased, for example. It may also be possible to reduce the variance, perhaps even to zero (at least as far as behavioural expression of criminal tendencies is concerned); I would personally doubt very much whether this possibility really exists in any serious sense, but this is a personal viewpoint and not relevant to the main bone of contention. The demonstration of partial hereditary determination of criminal conduct in our type of society says nothing about the possibility of creating other types of society in which criminal conduct might be very much reduced or even abolished; such possibilities must be evaluated in terms of other criteria. Marxists certainly have no quarrel with the facts on these grounds; our demonstration does not deny the possibility of their particular Utopia. It seemed worth while to point this out in some detail as the contrary point is presented with monotonous regularity by sociologists better acquainted with the writings of Marx than with those of modern geneticists.

We must next turn to a discussion of mental disorders, particularly those called 'neuroses' or 'psychoneuroses'. Here, the distinction between neurosis and psychosis may need some clarification. Psychosis is much the more severe form of mental disorder, more nearly akin to what the man in the street means by 'madness'. Schizophrenia and manic-depressive illness are the best known psychotic disorders; sufferers are likely to be segregated in hospitals, and until quite recently, when advances in drug therapy have made cures possible, the outlook was not encouraging. Neurosis is a much milder form of illness, usually; anxiety states, reactive depression, hysteria, obsessive-compulsive disorder are examples of neurotic states. Neurotics will usually be treated by their GPs (probably with tranquillizers), or go to the psychiatrist (privately or in hospital outpatient departments). Treatment will usually be by psychotherapy, or more recently by behaviour therapy, although drug treatment often accompanies this. Psychotic and neurotic disorders are quite different, nor does one lead to the other.

There are a number of concordance studies relating to neurotic disorders, as shown in Table 12, and these tend to demonstrate quite clearly that in every case identical twins show greater concordance than do fraternal twins. The total number of cases involved is rather smaller, unfortunately, so that our conclusions cannot be as certain as in the case of criminality; nevertheless, identical twins show concordance rates almost exactly twice those shown by fraternal twins. There are in all 141

Author:	Identical:	Fraternal:
Slater, 1953	2/6 = 25%	8/35 = 19%
Ihda, 1960	10/10 = 50%	2/3 = 40%
Braconi, 1961	18/2 = 90%	13/17 = 43%
Tienari, 1963	12/9 = 57%	not included
Parker, 1964, 1965	7/3 = 70%	4/7 = 30%
Shields & Slater, 1966	25/37 = 40%	13/71 = 15%
Total:	74/67 = 59%	40/133 = 30%

Table 12 *Concordance rates for neurosis: identical and fraternal twins*

identical twin pairs and 173 fraternal twin pairs, and these numbers are quite sufficient to assure the statistical significance of the results. The genetic component tends to be greatest in such types of neurosis as obsessional neurosis, anxiety states, and phobic complaints (the dysthymic neuroses); it tends to be least in such disorders as hysteria. It is interesting to note that here too specificity of disorder seems to be powerfully determined by genetic factors. In fact Shields concluded from his review of the literature that 'the particular combination of many genetic factors in the neurotic individual influences the form of his neurosis to a greater extent than it influences whether or how severely he breaks down.'

Another approach to the problem of the influence of genetic factors in neurosis is through the study of families of neurotic patients; it has been shown by several writers that there is an excess of neurotic relatives in the families of such patients. There is a history of neurotic illness in about 15% of parents and siblings of anxiety neurotics, a much higher percentage than is found in the families of controls. Here too there is a certain degree of specificity; relatives are not only concordant with respect to neurotic disorder, but also with respect to the particular symptomatology shown by the patient. Alcoholism might also be discussed in connection with neurosis, as it is certainly a non-psychotic behavioural disorder; there seems to be little doubt about an important genetic contribution to the position of a person on the abstinence—drinking—excess drinking—alcoholism dimension. A number of twin studies, mostly carried out in the Scandinavian countries, have shown that identical twins are far more concordant than are fraternal twins in this respect, and family studies have borne out this finding. Alcoholism seems to be inherited to the same extent as neurosis (and may of course be a reaction to neurotic stress).

Neurotic disorders are usually less debilitating than psychotic ones (particularly schizophrenia) and also far more widespread. A minimum estimate of the incidence of neurosis is that at some time of their lives, 10% of men and 20% of women in our culture will have a psychoneurotic attack; the true figures are probably higher than this because difficulties in diagnosis and lack of treatment facilities probably hide a large proportion of the true burden of this affliction. Neurotic patients are not usually hospitalized, whereas psychotics tend to be inpatients, although this pattern is changing a little now that drug treatments and other physical-type treatments have been found to be effective in producing a cure, or at least an amelioration, in psychotic illnesses. Psychotic patients occupy about half of all hospital beds in this country, or did so until quite recently; this was so because of the failure of traditional methods to touch the illness, making their stay in hospital very lengthy, rather than because of the large numbers of patients involved.

The incidence rate of schizophrenia is about 1%, but this is very dependent on social class; the rate is about three times as high for the lower social classes than for the highest ones. Manic-depressive illness, the other main functional psychosis (meaning a psychotic illness not produced by toxic or other known chemical or physical factors) shows a similar incidence rate, but is much less dependent on social class; indeed, claims have been made that it occurs more frequently in the upper classes. Although these two illnesses appear quite dis-

tinct when described in text-books, nevertheless diagnosis is not very reliable and is subject to national
differences; identical patients have a much higher chance
of being diagnosed schizophrenic in the USA than in
the UK, the probability of such a diagnosis being five
times as high in the one country as in the other!

What are the main indications and symptoms of
schizophrenia? Thought associations lose their con-
tinuity; there is a deterioration of attention; will and
initiative deteriorate; the patient lives in his own inner
world and becomes detached from reality; there is considerable emotional deterioration; ambivalence is pro-
minent, that is the presence of opposing or contrary
feelings, desires, or thoughts; there is inadequate adap-
tation to the environment, accompanied by confusion,
sudden fantasies or peculiarities, and disregard of
reality. Furthermore, there is a type of 'dementia',
which shows itself in numerous silly mistakes in thinking
or acting, or by stupidity and foolishness in spite of
sometimes adequate performance on, say, IQ tests and
other measures of mental ability. These symptoms may
either appear in a very severe form, usually in deteriorated and hospitalized patients, and certain psychiatrists
would only diagnose schizophrenia in these cases; or
else they may appear in a relatively mild form, shading
over into types of odd behaviour which can be observed
in many otherwise seemingly ordinary people who
clearly do not merit the diagnosis of 'schizophrenia'.
This obviously presents the investigator with many
difficulties; in twin studies, for instance, can we call a
case 'concordant' where the proband is definitely
schizophrenic according to the first meaning of the
term, while his co-twin is merely 'schizoid', that is extremely odd in his behaviour, but not definitely certi-
fiable? Other difficulties arise because there are various
sub-types of schizophrenia, such as catatonia, paranoia,
hebephrenia, etcetera; again these tend to shade into
each other, and diagnosis is by no means very reliable.

Manic-depressive psychosis is more or less self-descriptive. It involves alternating periods of severe depression and mania, although for many people the alternation is merely from depression to normality and back, or occasionally from mania to normality. Mania manifests itself in a sense of elation, a flood of ideas that pours out into a stream of speech that may become incoherent. Irritability and excitement predominate as far as mood is concerned, and the patient may be very lively and energetic, or irritable and impatient. In the depressive state the patient is listless, indifferent, sluggish, tearful, and shows feelings of unworthiness, despair, and of being evil. Delusions are also often present, either in the manic or the depressed state; these are usually related to the patient's mood—thus he may feel that his internal organs are rotting. These attacks tend to be self-terminating, that is, the patient gets better without any treatment, although now there are drugs and other types of physical treatment which accelerate recovery.

Psychotic disorders are quite distinct from neurotic ones, although psychoanalysts used to think differently—they considered psychosis as merely a more severe form of neurosis. This is almost certainly untrue. There are certain biochemical peculiarities in schizophrenics which are not found in neurotics; there are differences in neural reactivity which are only found in schizophrenics, not in neurotics; cortical evoked potentials show quite distinct recovery patterns in psychotics and in neurotics; sedation thresholds differ profoundly between neurotic and psychotic depressions; patterns of onset are quite different, particularly between schizophrenia and neurosis; psychological tests show entirely different types of reaction for neurotics and for psychotics. These and many other reasons, including differential response to psychotherapy and drug therapy, make it almost certain that psychoses and neuroses are quite distinct disorders, and that no useful purpose would be served in throwing them together.

Schizophrenia is powerfully determined by genetic causes, although not to the exclusion of environmental ones. There are many lines of evidence clearly pointing to this conclusion. The first is the markedly increased familial morbidity risk; the risk of becoming schizophrenic is much greater if there are other schizophrenics in the family. Siblings of schizophrenics have a risk of becoming schizophrenic which is ten times as high as that existing in the general population. Similarly for other degrees of relationship: the risk of schizophrenia is always higher in relatives of schizophrenics, however distant. Nor is this explained by saying that it is the fact of being brought up by schizophrenic parents, or in contact with schizophrenic brothers and sisters, which produces the disorder. Newly-born babies have been taken from schizophrenic mothers immediately after birth and farmed out to foster parents who were perfectly normal; yet the incidence of schizophrenia in these children proved to be many times that in the normal population—in spite of their being brought up in a perfectly normal environment. The morbidity risk of the children of schizophrenic parents is about the same as that of siblings, that is about one chance in ten; being brought up by normal foster parents apparentyl does little to change this figure.

Twin studies give results which firmly support this conclusion; unfortunately these results cannot easily be tabulated like those for crime and neurosis because authors usually give different percentages for the same population studied, depending on the precise criteria employed. But revision of the concordance rate for identical twins also implies a similar revision in the concordance rate for fraternal twins; by and large the best way of reporting the data is perhaps by averaging both the highest and the lowest estimate in each case where there is more than one estimate. Some five hundred pairs of identical twins, and one thousand two hundred pairs of fraternal twins have been studied; the concordance rates are 35% or 53% for the identicals, depending on diagnostic criteria, and 7% or 10% for the fraternals. Whichever way we look at these figures, that for identicals is five times as great as that for fraternals, which means that the difference between one kind of twin and the other for concordance is much greater even than it was for neurosis or crime. Nevertheless, the concordance rate for identicals is well below the 100% mark, which indicates that nongenetic factors have an important role to play with respect to the development of schizophrenia.

These of course are twins reared together; what about identical twins reared apart? Only sixteen such pairs have been reported; of these, ten were concordant, giving a rate of 62.5%, which is actually higher than the rate for identical twins reared together! This should not be interpreted as proving that environmental factors that make for discordance have nothing to do with rearing practices; there is always the possibility that sampling errors may have biased the result. Nevertheless, the result is interesting, and must be given some weight in company with all the other evidence. As always, it is not proper to rely entirely on one kind of evidence in coming to a conclusion; what makes a theory acceptable is that many different lines of evidence point to the same conclusion.

Another way of attacking the problem is by looking at children offered for adoption, and comparing two samples—one consisting of children who grew up to become schizophrenic, the other of children who showed no sign whatever of schizophrenia; these groups were of course matched for sex, age, and age of transfer to the adopting family. All adoptive and biological parents, siblings and half-siblings of both groups were then identified, and an attempt made to see whether any of these persons was schizophrenic. Environmentalists would expect the relatives of adoptive families to show a greater number of schizophrenics, while interactionists would expect the relatives of the biological families to contain more schizophrenics. The outcome was clear. There was a significantly greater number of schizophrenic-type disorders among the *biological* relatives of the schizophrenic index cases than among the relatives of the controls. Among adoptive relatives, however, there was no difference between the two groups. This study, which supports the genetic-interactionist hypothesis, was carried out in Denmark; a similar study done in Iceland corroborates it. In this study, the biological and foster siblings of schizophrenics who had been adopted before the age of one were compared; among the former, six out of twenty-nine were schizophrenics, while among the latter, none of the twenty-eight were schizophrenic. These figures are very convincing in demonstrating the importance of the biological, as opposed to the environmental influences (see Shields, 1973).

If there is a strong biological, genetic predisposition towards schizophrenia, then one would expect to be able to discover specific genetic markers for the disease. Recent research has examined the enzyme content of platelets in the blood of schizophrenics. It was found that monoamine oxidase (MAO) is present in very reduced quantity in the platelets of schizophrenics as compared with normals; moreover, in the identical co-twins of schizophrenics lowered MAO levels have also been found, suggesting that this reduction is an index of the predisposition to the disease, and not a consequence of it.

Investigations have not only established that heredity plays a very powerful role in the genesis of schizophrenia; they have further demonstrated that the sub-types of schizophrenia (hebephrenia, paranoia, catatonia) also tend to be inherited to some extent as separate units, although not sufficiently so as to make the concept of schizophrenia as a unitary disorder superfluous. Identical twins never have different subtypes, but within the same family different subtypes do occur. The most parsimonious hypothesis which has been put forward to explain all the facts seems to be that schizophrenia is inherited polygenically, that is through a number of genes rather than through a single gene, and that further the pathological genes tend to cluster more in certain families than in others: those with most of these genes will show hebephrenia and/or catatonia, while those with fewer of the genes will show paranoid, simple, atypical or even borderline forms of the disorder. To which one might add that with still fewer malignant genes we would presumably get the various oddities of human behaviour into which schizophreniform disorders tend to shade, and which are sometimes called 'schizoid'. This model is sometimes referred to as the 'diathesis-stress' model; diathesis means that there is a biological predisposition to develop schizophreniform disorders, of varying strength in different people, while stress is needed to produce the actual disorder and make it flourish. Thus diathesis refers to the genetic component, stress to the environmental one. This theory is now widely accepted, particularly since alternative hypotheses, such as single-gene theories, or purely environmental theories, have been shown to be wrong.

Findings with respect to manic-depressive disorders are in most respects very similar to those discussed with respect to schizophrenia, and no lengthy discussion will be necessary. There is again the strong presence of a 'familial taint', and there is again the evidence from twin studies that identicals are far more concordant than are fraternals; the figures are around 70% for identicals and around 18% for fraternals. It seems possible that the cyclic form of the disorder (from mania to depression, and back) is more strongly inherited than are the monopolar forms (only mania or only depression, veering to normal and back). Here, as in the case of schizophrenia, we may posit a diathesis-stress model, but it should be added that in neither case do we really know very much about the 'stress' part of the equation; there have not yet been the properly controlled and designed studies which alone can give us information on the vital question of just what it is in the environment which produces the stress which triggers off the illness. Clearly it is not just any stress; studies have been done into the effects of stress produced by war-time service in the armed forces, with the result that duration of service produces a greater and greater incidence of neurotic illnesses, but fails to increase the incidence of psychotic disorders. To postulate environmental causes tells us nothing; this is implicit in the distinction between geno-type and phenotype. Our knowledge would only be increased if we could be shown what it is in the environ-ment that works together with the hereditary predis-position to produce the psychotic breakdown. This we do not know, and there are not even any good theories at present which could stimulate research.

Is there any genetic link between schizophrenia and manic-depressive illness? The general tenor of the litera-ture seems to be in the negative, but there are powerful reasons for believing that this conclusion may be mis-taken. In a recent study Schulz (1951) found that among the offspring of parents both of whom were manic-depressive, the incidence of manic-depressive disorder was 28% but that for schizophrenia was 12%, which is of course very high—much higher than the in-cidence in the population. Quite generally, there is evidence to show elevated risk of schizophrenia in the children of manic-depressive parents. Further, on psycho-logical tests manic-depressive and schizophrenic patients are difficult to distinguish, although both groups per-form quite differently to neurotics. Last, if these two groups of disorders were clearly distinguished on a hereditary basis, then it should be relatively easy to diagnose a patient as falling into one or the other; this, as we have seen, is not so. Diagnoses are very unreliable (except in the rare 'text-book' cases) and indeed may differ from country to country; this does not suggest a fundamental division. (An interesting suggestion that different types of psychotic illness may have in common some bond comes from work on season of birth. Seven per cent more schizophrenics and 9% more manic-depressives were found to be born between the end of December and the beginning of April in one Maudsley study, thus placing their favoured conception time between April and June. It is not clear why this should be so, but the fact remains that both types of psychosis are equally involved in this odd phenomenon, suggesting that they have something of a biological nature in common.)

My own conclusion would be that we may be justified in postulating a hierarchical scheme, similar in kind to that we found suitable for intelligence. At the top of the hierarchy we have a general predisposition to psychosis, which by analogy we may perhaps call 'psychoticism'; at a slightly lower level we have two major 'group factors', schizophrenia and manic-depressive illness. These can then be subdivided again into 'primary factors' such as hebephrenia, paranoia, catatonia, etcetera. General 'psychoticism' is strongly determined genetically, but the more specific manifestations at the lower levels also have specific genetic determinants additional to those which go to produce 'P' or general psychoticism. This scheme accounts for nearly all the known facts, but it must be admitted that other schemes are equally feasible, and that conclusive evidence is lacking.

We must now turn to another group of mentally abnormal individuals who provide a link between criminals, with whom we started this discussion, and schizophrenics. These are the so-called 'psychopaths', defined as follows by the *Diagnostic and Statistical Manual of Mental Disorders* of the American Psychiatric Association: '. . . chronically antisocial individuals who are always in trouble, profiting neither from experience nor punishment, and maintaining no real loyalties to any person, group or code. They are frequently callous and hedonistic, showing marked emotional immaturity, with lack of responsibility, lack of judgment, and an ability to rationalize their behaviour so that it appears warranted, reasonable and justified.' Others describe the psychopath as being characterized by superficial charm, unreliability, untruthfulness and insincerity, lack of remorse or shame, antisocial behaviour without good cause, poor judgment and failure to learn from experience, pathological egocentricity and incapacity for love, unresponsiveness in general personal relations, impersonal, trivial and poorly integrated sex life, and lack of any real life plan. Individuals of this kind may or may not be criminals, and criminals may or may not be psychopaths. There is much overlap, but nothing like identity.

Frequently the psychopaths as described above are called 'primary', in contradistinction to another type of person labelled neurotic or 'secondary' psychopath. Many antisocial and aggressive acts are committed by people who suffer from severe emotional disturbances, unbearable frustrations and inner conflicts. Their aggressive and antisocial behaviour is believed to be the consequence of more basic emotional problems. Secondary psychopaths will be discussed later on; here we will rather draw attention to the close relation between psychosis and primary psychopathy. This emerges from many studies of the progeny of psychotic (mainly schizophrenic) parents. While these, as we have seen, tend to contain a large number of psychotics, they also invariably contain a possibly larger number of criminals and primary psychopaths. (Often the term 'schizoid psychopaths' has been used to characterize these psychopathic offspring of schizophrenic parents.) We have already noted a study in which the new-born babies of schizophrenic mothers were taken away from them at birth and handed over to foster parents; in addition to the large proportion of these babies who developed schizophrenia when adult, an even larger number developed psychopathic conditions or became criminals. Thus there is a close genetic connection between primary psychopathy and psychosis; just as there is a connection between secondary psychopathy and neurosis. This point will be seen to be important when we turn to the discussion of the inheritance of personality.

Before turning to this discussion, we may perhaps note briefly how very different are the accounts of the origin of schizophrenic and other psychotic illnesses here given from those offered by such purely environmentalistic writers as R. D. Laing, D. G. Cooper, and T. S. Szasz. The views of Laing, in particular, seem to lay all the blame for schizophrenic types of behaviour on the parents of the patients. Thus, in one book he concludes a case history with the statement that the patient's numerous and serious symptoms are '. . . the outcome of her interexperience and interaction with her parents'. Certainly the causal role of the family in producing schizophrenia is widely thought to be the main contribution of Laing and his school to the understanding of this disorder, although he himself, in the book *Sanity, Madness and the Family*, argues that criticism of this view 'would be justified if we had set out to test the hypothesis that the family is a pathogenic variable in the genesis of schizophrenia. But we did not set out to do this, and we have not claimed to have done so.' Laing seems to claim rather to have succeeded in making schizophrenic behaviour 'intelligible', although this claim too may be doubted. However successful he may sometimes be in making the patient's impoverished emotional responses meaningful, he never gives an intelligible account of such crucial schizophrenic phenomena as hallucinations, or catatonic immobility, or prolonged apathy, etcetera. And in most of the instances cited, there is a huge disproportion between the apparently damaging behaviour of the relatives and the duration, extent and depth of the patient's handicaps. In many cases the family behaviour which is said to make the abnormal behaviour of the patient intelligible seems to be quite trivial and no different from behaviour seen in many situations where there is no question of schizophrenia resulting. As Rachman (1973) has pointed out in his critique:

> it is hard to avoid the impression that they have set out to demonstrate that the exceedingly abnormal behaviour of the patients is a direct reflection of the equal or greater disturbance in the family members and to this end, they mix description with inference and inference with interpretation. Their evident sympathy for the patients does not extend to the relatives who are described and discussed with little sign of compassion. In all it might have seemed more convincing if they had encountered at least some parents who were not completely devoid of insight or genuine affection.

Laing and his followers never consider the contrary evidence presented by such studies as those in which children of schizophrenic mothers develop schizophrenia in spite of being brought up from early babyhood by foster-parents whose behaviour is quite non-pathological. They never consider the possibility that abnormal behaviour in parents may be a direct consequence of having a pathological child; parental attitudes of the mothers of schizophrenic children were found to be pathological in one study, but less so than the attitudes of parents of brain-damaged children! And of course Laing and his followers never consider the genetic evidence, which puts the major blame for the disease squarely on the genes, rather than on the behaviour of the parents. It is not unlikely that in certain cases familial stresses may precipitate a schizophrenic disorder in an individual genetically predisposed to this disorder; yet even this has not been adequately proved. To go beyond this and claim that this is the only factor which, in the *absence* of genetic predisposition, could produce these results, is to abandon science for mysticism.

In turning from a consideration of socially important concepts such as criminality, neurosis, or psychosis to personality variables such as emotionality or extroversion, we may seem to be abandoning one set of concepts and going over to another. We shall soon see that this is not in fact so, and that there is indeed a close relation between the socially important variables we have considered so far, and the personality variables to which we turn now. It would take us too far afield to discuss in detail the many meanings of the term 'personality'. We shall here use the term to mean semi-permanent patterns of behaviour characteristic of individuals which are of social importance and relevance. The notion of individual differences is obviously central to the concept of personality, but there are many accidental and unimportant differences between individuals which would not be considered part of personality by most psychologists. The most widely accepted conceptualization of these individual differences is in terms of traits and types, concepts analogous to the notion of 'ability' in the cognitive field, and there is now considerable agreement on some form of hierarchical scheme, in which 'type' concepts, such as extroversion, are based on the observed correlations between traits, such as sociability, impulsiveness, activity, liveliness, excitability, etcetera. Figure 10 shows such a hierarchical scheme in outline. Traits themselves are similarly based on observed correlations between different manifestions; thus the trait 'sociability' is based on the observation that individuals who like to go to parties also like to have many friends, like 'to talk to strangers', and dislike reading in solitude. Each of the items making up a trait taps habitual response patterns; for example, liking to talk to strangers usually, not just on the odd occasion. Some responses are, of course, quite specific, but these do not really concern us very much in constructing a model of personality.

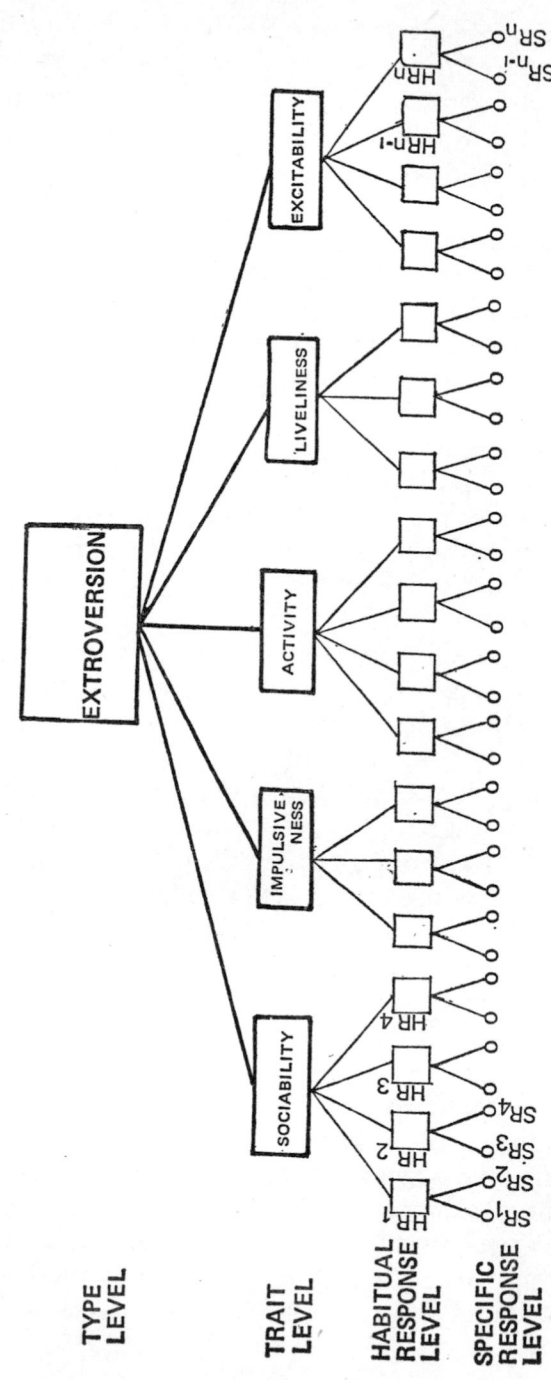

Figure 10 *The hierarchical conception of personality*

Type concepts are not differentiated from trait concepts in terms of distribution. Just as sociability is fairly normally distributed, with most people around the average mark, and fewer and fewer extremely sociable or extremely unsociable, so is extroversion. The erroneous notion that type concepts imply bimodal distributions, or actual categorical divisions into mutually exclusive and qualitatively different 'humours', dates back to the ancient Greeks and does not form part of the teaching of modern psychology. Yet there was some degree of truth in these older notions. Most modern studies of personality have come up with two major 'types' or dimensions of personality: the first of these refers to the opposition between extroverted and introverted personalities, the other to the opposition between people with strong, labile emotions and people with less strong, stable emotions. This latter dimension is often referred to as one of 'emotionality', 'anxiety', or neuroticism. The letters E and N are frequently used as a shorthand notation to denote these two dimensions: extroversion and neuroticism. Figure 11 shows how these dimensions are related to the four Greek 'humours' or types; melancholics and phlegmatics are introverted, cholerics and sanguinics are extroverted. Similarly, melancholics and cholerics are emotionally labile, while phlegmatics and sanguinics are stable. The trait names inside the circle may serve to give an idea of the behaviour patterns characteristic of extroverts and introverts, labile and stable people—remembering always that extremes in either direction are rare, and that most people are somewhere intermediate.

There is a direct connection between these personality dimensions and the study of neurosis and criminality. We will first consider the descriptive connection, then turn to the causal or explanatory aspects. Let us note, then, that many studies have shown that neurotic patients tend overwhelmingly to fall into the quadrant which combines high emotionality with strong intro-

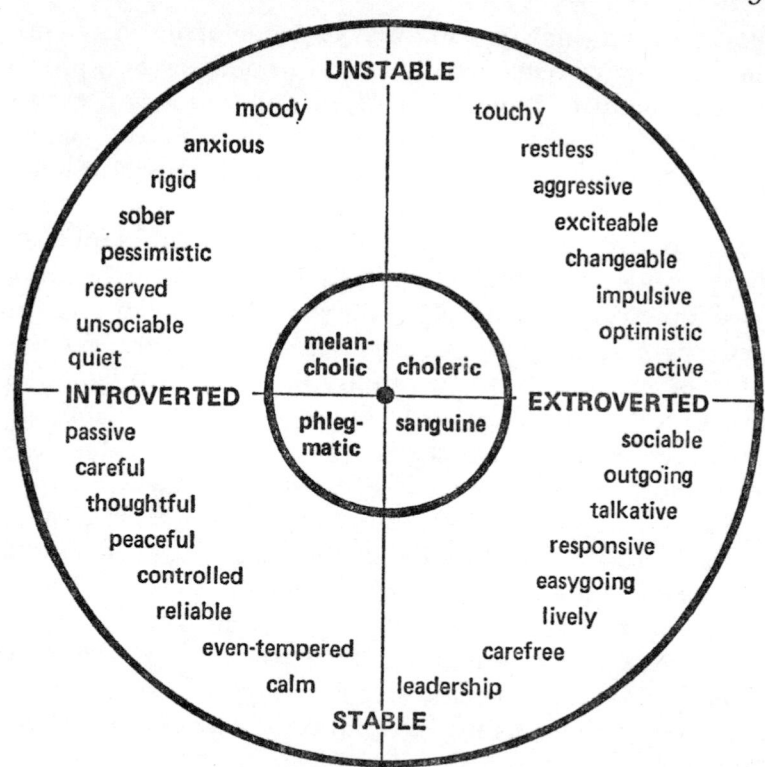

Figure 11 *The relation between the theory of the four temperaments and modern dimensional theories of personality*

version; this corresponds to the 'melancholic' group in Figure 11. Similarly, criminals tend to fall into the 'choleric' quadrant, that is they combine high emotionality with strong extroversion. In view of the marked heterogeneity of 'criminals' it must be obvious that there are many exceptions to this rule; nevertheless, many studies, in various different countries, have shown that on the whole the number of criminals falling into this quadrant is significantly greater than would be found in a random sample of the population, or even in one matched with the criminals on age, sex, and social status. Along the same lines, follow-up studies of children rated on scales of E and N show that they tend to develop very much in a manner expected from this hypothesis; those who eventually become confirmed criminals tend strongly to have been rated emotional and extroverted as children, while those who eventually become neurotics tend strongly to have been rated emotional and introverted as children. In terms of our discussion of psychopathy, we might say that this combination of high emotionality and extroversion makes for secondary psychopathy; we shall later on see what makes for primary psychopathy. Let us only note here that our dimensions of personality fit in remarkably well with some of the socially important groupings (neurotics, criminals) we started out with; this suggests that personality plays an important part in causing these types of behaviour.

These factors or dimensions are purely descriptive; it seems reasonable to seek for a causal, physiological basis for the different types of behaviour which we call extroverted or introverted, emotional or stable. I have suggested such a causal model in my book on *The Biological Basis of Personality* (1967); this is not the place to go into detail regarding this model, and only a brief outline will be given.

It is suggested that introversion is produced essentially by high arousal levels in the cortex; this high arousal level, in turn, is produced by an over-active ascending reticular formation. Figure 12 gives a rough diagrammatic picture of the situation. Incoming sensory messages from the outer world go directly to the cortex, but they also send collaterals into the ascending reticular activating system, which in turn then bombards the cortex with a stream of signals the primary purpose of which is to keep it awake and aroused so that it can deal with the incoming messages from the outer world. Without this bombardment the cortex would not be able to respond, and would be in a state of quiescence.

An aroused cortex possesses certain characteristics, as compared with an underaroused one, which may be useful in explaining the behavioural peculiarities of ex-

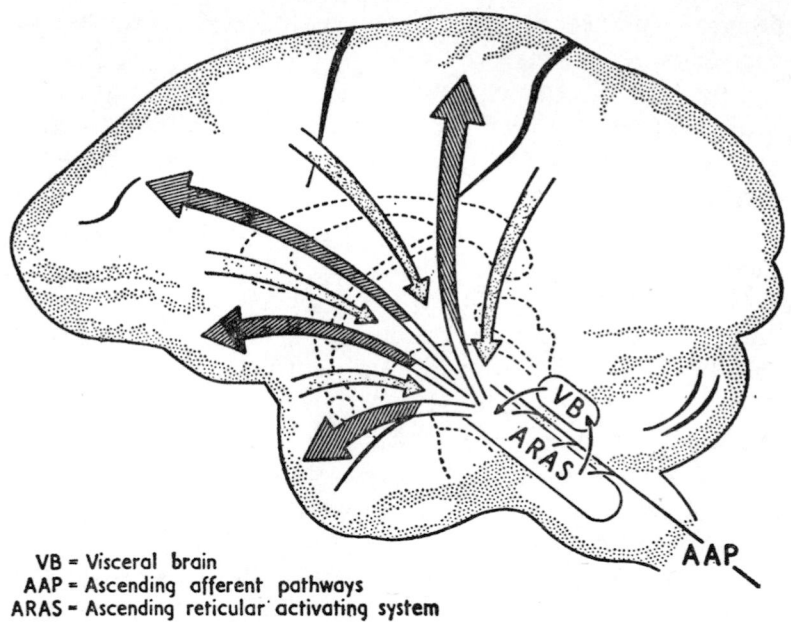

VB = Visceral brain
AAP = Ascending afferent pathways
ARAS = Ascending reticular activating system

Figure 12 *Diagrammatic view of physiological structures underlying main dimensions of personality*

troverted and introverted persons. Arousal helps an individual to learn, and to form conditioned responses; hence we smoke and drink coffee when we have to keep alert during a period of study, and drink alcohol to relax. (Nicotine and caffeine are stimulant drugs, producing increased arousal, while alcohol is a depressant drug, having the opposite effect.) This greater ease of learning and conditioning may be responsible for the greater ease of 'socialization' of introverts; obviously some form of social learning is involved in producing the socially acceptable behaviour which keeps us out of prison. (I have spelled out in some detail just how this process may be conceptualized in my book on *Crime and Personality* (1970) and will not here repeat what I have said there.) Neurotics, on the other hand, form conditioned patterns of anxiety and other inappropriate emotional responses all too easily, being introverted by

7

nature; this accounts for the prevalence of introverts in the neurotic group.

There is another charâcteristic of the poorly aroused brain of the extrovert. People have a preferred, median level of arousal—if this is too low, they are bored, while if it is too high they are upset. Extroverts tend to have a level which is too low much of the time, unless environment can provide excitement and stimulation; hence they tend to be stimulus-hungry and sensation-seeking. This sensation-seeking proclivity is very likely to be closely related to criminal conduct; many future criminals start out by being attracted to the bright lights and the loud pop music of city life. We thus have, as a consequence of low arousal levels in the cortex, greater temptation (sensation-seeking behaviour) and less resistance to temptation (poorer level of socialization); when these characteristics occur in conjunction with strong emotions (as in the 'choleric' type) then it is not unreasonable to expect criminal behaviour to occur with greater than chance probability.

Emotionality or neuroticism also has a close connection with one of the great bodily systems. All forms of emotional expression and behaviour are intimately linked with the so-called autonomic system, which governs such 'automatic' responses as heart-beat, breathing, blood-flow, digestion, etcetera. Most people will recall that in fear or anger it is these physiological reactions which are most noticeable—the rapid beat of the heart (in order to make more blood flow through the body), the rapid breathing (to make more oxygen available), the cessation of digestion (to divert blood from the stomach), and other adjustments to prepare the organism for fight or flight. These and many other, less noticeable reactions of the body constitute the basis of our emotional feelings, and they tend to be more labile and easily aroused (and less easily stopped) in highly emotional people. Emotional reactions are co-ordinated and regulated by the so-called visceral brain (see Fig. 12) and it is here that one would seek the main physiological habitat of the personality dimension 'emotionality' or neuroticism.

There is much direct physiological work implicating these various systems with E and N; thus for instance studies of the electroencephalograph have shown that introverts tend to have low amplitude and high frequency alpha waves, typical of high arousal, while extroverts tend to have high amplitude and low frequency alpha waves, indicative of low arousal. Similarly, high N and low N subjects tend to show quite different physiological reactions to emotion-producing stimuli in the laboratory. More indirectly, traditional experimental laboratory studies of conditioning, vigilance, sensory thresholds, learning, and other psychological functions have given strong support to some such system as that outlined above. Clearly the model is oversimplified and far from complete; nevertheless it does seem to have some application to the problems with which we are here concerned.

The model becomes a little less restricted when we add another major dimension, namely that of psychoticism (P). We have already seen that psychotic illnesses shade over insensibly into schizoid behaviour, and also into odd but not pathological behaviour such as we might see every day; this suggests that underlying the florid symptomatology of the psychotic there is a particular personality structure which can possibly also be identified in 'normal' people. There is direct evidence for the hypothesis that psychotic illnesses represent the extreme of a continuum, the other parts of which are filled with non-psychotic individuals, just as neurosis too is merely the extreme of a continuous distribution.

Based on this reasoning, a questionnaire was constructed which fulfilled the following conditions: (1) It should differentiate psychotics from normal and from neurotic subjects, with psychotics having particularly high scores. (2) It should produce higher scores for more deteriorated psychotics than the less severely ill psychotics. (3) It should show a decrease in scores for psychotics who had (partially or entirely) recovered from their illness. (4) The different items in the questionnaire should show the same pattern of intercorrelations for a normal as for a psychotic sample. (5) Scores on the questionnaire should correlate with ratings of severity of illness in psychotic patients. (6) Scores on the questionnaire should correlate with performance on psychological tests known to measure deterioration of psychotic patients. (7) Scores in normal subjects should correlate with performance on psychological performance tests known to differentiate between psychotics and normals. Such a test, fulfilling all these conditions, has now been in use for six years or so, and it is interesting to note that this test also fulfilled a further condition that was originally considered a reasonable prerequisite, namely that the test should discriminate between criminals and non-criminals.

The rationale of this added requirement will be obvious: if primary psychopathy is genetically linked with psychoticism, as we have pointed out above, and if psychopathy leads to criminal conduct in many cases, then a P scale, measuring psychoticism and psychopathy in combination, should elicit higher scores in prisoners (at least in those with primary psychopathy) than in non-prisoners. Several thousand prisoners have now been tested, and the predicted difference has been found in every study; it obtains just as clearly among women as among men. We may therefore feel that this scale does to some degree measure what it purports to measure, namely an underlying personality dimension which mediates psychotic and psychopathic conduct. It is of interest to note that this P scale elicits particularly high scores among criminals who are in prison because of aggressive and sexual offences, that is precisely those offences which would be expected to be committed by (primary) psychopaths.

The P scale was difficult to construct because what is most noticeable in psychotics is obviously the assortment of pathological symptoms which is so characteristic of the various types of psychosis. In making up a questionnaire of P for normal people such obvious symptoms were of course irrelevant, because normal people simply never show them. The aim was to look at the underlying personality features, which might also be found in 'normal' people.

These, then, are the traits which characterize high P scorers: (1) Solitary, not caring for people; (2) Troublesome, not fitting in; (3) Cruel, inhumane; (4) Lack of feeling, insensitive; (5) Sensation-seeking, underaroused; (6) Hostile to others, aggressive; (7) Liking for odd, unusual things; (8) Disregard for danger, foolhardy; (9) Liking to make fools of other people, upsetting them; (10) Opposed to accepted social customs, e.g. marriage. (11) Little personal interaction; prefers 'impersonal' sex. It is noteworthy that this scale does not correlate with either E or N; thus all combinations of these three major personality dimensions can be found, and of course modify each other. Just as a high N scorer may behave quite differently, depending on his E score, similarly, high P scorers behave quite differently depending on their E or N scores.

We can now turn to the question of the degree to which P, E and N differences between people are caused by genetic factors. Most textbooks of psychology seem to agree that the answer is in the negative; genetic causes in this field are not fashionable. This belief originated in the first place because of the outcome of a very famous study on identical and fraternal twins, carried out in 1937 by Newman, Freeman and Holzinger, and still widely quoted. These writers compared fifty pairs of identical and fifty pairs of fraternal twins with respect to a number of physical measurements, mental and educational tests, and personality tests; they concluded that 'the physical characteristics are least affected by the environment, that intelligence is affected more; educational achievement still more; and personality or temperament, if our tests can be relied upon, the most.' The lack of hereditary influence on personality here suggested is still assumed to be a true statement of fact by many psychologists who have not kept up with recent developments in this field, but there are reasons for regarding it with suspicion because of several weaknesses in the original study itself. Note the admirable caution with which the authors themselves hedge their bets by saying that these conclusions with respect to personality are true 'if our tests can be relied upon'; in fact there were no proper tests of personality available at the time, and thus those used by them could not in the nature of things give results of any value. Furthermore, the tests used had been standardized on adults, but the mean age of the twins was only thirteen; thus the tests were inappropriate. And in the third place, the conclusion cited does not even follow from the actual data given. The authors used a neuroticism questionnaire which, while not properly appropriate to children, might nevertheless be expected to give useful data. On this questionnaire, identical twins correlate .56, fraternals .37; identical twins brought up in separation correlate .58! These figures, particularly the high value for twins brought up in separation, suggest quite a sizeable contribution by heredity, particularly when the resulting heritability is corrected for the unreliability of the test, as it should be.

It is curious that identical twins brought up in separation should be more alike than identical twins brought up together, and the difference is certainly not large enough to build any conclusions on it; nevertheless, we shall see soon that other investigators have found similar results. The answer may lie in the often-noted tendency of identical twins to grow tired of their lack of independent identity, and to attempt to 'grow away from each other' by purposely developing different and compensating behaviours—thus one might become more sociable and deal with relations with other people, while the other twin might concentrate his interest on household affairs. This attempt to become different would not be present in identical twins brought up in separation.

However this might be, very numerous studies of hundreds of pairs of twins, both in the USA and in this country, have demonstrated beyond any possibility of doubt that many different types of personality traits and in particular extroversion and neuroticism, show much higher correlations in identical than in fraternal twins; the uncorrected heritability estimates tend to be around the 50% mark. None of these many studies show the high degree of competence evinced in an exceptionally important experiment reported by J. Shields (1962) in his book, *Monozygotic Twins*, and hence we may perhaps be excused if we concentrate on this one study, rather than discuss the many less important ones. Using a personality inventory measuring E and N, Shields studied forty-four separated pairs of identical twins, forty-four non-separated pairs of identical twins, and thirty-two pairs of fraternal twins of which eleven had been brought up apart. For extroversion, the identical twins brought up together and in separation correlated respectively .42 and .61; the fraternal twins correlated —.17. For neuroticism, the figures for the identical twins were .38 and .53, and for the fraternal twins the correlation was .11. Note again that identical twins are more alike when they have been brought up apart than when they have been brought up together.

These data were reanalysed by J. Jinks and D. W. Fulker (1970), using the most recently developed methods of biometrical genetical analysis. They found no evidence for genotype-environment interaction, nor for assortative mating; furthermore, there was an absence of dominant gene action. This, as they point out, strongly suggests that intermediate degrees of extroversion and neuroticism have been favoured by natural selection and constitute the population optimum for these personality traits. The heritability for neuroticism is calculated as .54, that for extroversion as .67; these are of course minimum values as no correction has been made for unreliability of the measuring instruments. Jinks and Fulker also found that 'introvert genotypes are more susceptible to environmental influences than extrovert genotypes, the latter being relatively impervious. This finding is, of course, fully consistent with Eysenck's theory that the introvert is more conditionable than the extrovert.' These findings then fully agree with the kind of psychological analysis we have been putting forward, making the poor response of the extrovert to environmental influences responsible for the prevalence of extroverts among criminals, and the over-strong response of the introvert to such influences for the prevalence of introverts among neurotics. However that might be, there can be no doubt about the importance of genetic factors in causing personality differences in E and N.

As far as P is concerned, only one study (Eysenck, 1972) has been reported in which this scale was administered to forty-five pairs of identical twins living together and fifty-seven pairs of identical twins living apart; these twins were also administered tests of E and N. Heritability estimates were derived from the scores obtained, and it was found that P had a heritability index of .53; the respective indices for E and N were .37 and .65. (These figures are less accurate than those from the Shields study, and should not be overinterpreted.) There is no doubt that P, too, is determined by hereditary causes to a considerable extent, but there is not yet enough evidence to be certain of the degree to which this is true, or whether P differs in any way from E and N in the degree of heritability. It remains to be said that all these estimates of heritability would rise dramatically if the appropriate corrections for unreliability of the personality scales were made. Such corrections have not been attempted here as the level of accuracy possible with the relatively small numbers of twins studied by different investigators is not high enough to give a really convincing estimate of the 'true' heritability. Work is under way to remedy this fault; for the purpose of this book no very refined estimate is in fact required as long as it is agreed that the evidence points quite strongly to an involvement of heredity in the causation of personality differences in P, E and N which certainly accounts for not less than 50% of the total variance, and may account for as much as 70%. This is a conservative conclusion; the true figures may be somewhat in excess even of the higher figure quoted.

We may conclude, therefore, that however we look at the facts, that is, whether we start from socially defined behaviour patterns (crime, neurosis, psychosis) or whether we start from psychologically measured dimensions of personality, heredity plays a very important part in the causation of these many different types of conduct and behaviour, and is responsible for a good proportion of the individual differences observed in our type of society. If the involvement of heredity in these aspects of behaviour seems somewhat less than in the case of cognitive abilities, this may be due in part to the unreliability of our measures, and in part to their crudity; lack of validity requires correction as much as lack of reliability. If we had better measures, it does not seem unreasonable to expect heritability estimates for personality dimensions similar to those found in the ability field.

Before closing this chapter, we must turn to one other topic which is closely related to personality, namely that of sex differences. In the well-known story of the suffragette who closed her speech by saying that there was only a little difference between men and women, a heckler rose to shout: 'Vive la petite difference!' He little realized that in fact he was opposing what was to become a platform of one wing of 'women's lib'—the belief that all observed differences between the sexes, except those concerned with primary and secondary sexual characteristics, are due to cultural rather than biological causes. This belief will not stand up to scrutiny, even if we agree to call differences in strength and agility 'secondary sexual characteristics'; the reality of these differences will be obvious to anybody who has watched sport and knows the huge differences in standard between men and women in tennis, squash, and even table tennis—let alone boxing, wrestling, and football! There certainly are tremendous differences between the sexes in almost every respect except intelligence—and even there, as we shall see, men and women tend to differ, although not very strongly, with respect to their special areas of strength and weakness. Psychologists have typically shied away from stressing these differences: experimenters often fail to report sex differences in their studies, or gloss over them when they do report them. This curious prejudice has even entered the field of rat work. A thoroughgoing belief in 'equality' has often blinded experimenters to the reality of sex differences in rodents! Needless to say, the fact of sex differences does not alter in any way the proposition that both sexes should and must have equal legal and other rights: it merely means that the sexes are complementary in many ways, rather than competitive. But to say all this is to prejudge the conclusion. Let us look at the evidence.

The first point to notice is that with respect to personality women have significantly higher scores than men on introversion and neuroticism, but lower scores on P. Descriptively they thus emerge as more fearful and emotional, less risk-taking and outgoing, less aggressive and hostile. These descriptions will be readily recognized as agreeing well with widely-held stereotypes, and they are in good accord with the sort of sex differences which one would expect evolutionary processes to produce. The childbearing and physically weaker sex, in order to survive and procreate, would have to develop habits of avoiding rather than seeking combat, of caring for the babies while the stronger, non-childbearing male would go out into a hostile world to forage and fight. Can we account for these widespread and almost universal differences simply in terms of cultural norms, or do these patterns of behaviour have more deep-seated and biological roots? There are two lines of evidence which compel an answer in terms of genetic differences. In the first place, these sex differences can be observed, not only in man where cultural effects might be presumed to have some importance, but also in animals (particularly mammals of course) where the very notion of cultural effects is meaningless. And in the second place, we can tie in aggressive and fearful behaviour with hormonal secretions which in turn are closely linked with the sexual apparatus. Together, these two lines of evidence leave very little doubt about the biological origin of these sex-linked differences in behaviour and personality.

Let us consider first of all aggressiveness. In man, the evidence is overwhelming that aggressive crimes are almost entirely carried out by males. In children, the tendency towards greater rough-and-tumble play in boys as compared with girls can be seen as early as the three to five age range. Non-human primates, like the macaque, the rhesus monkey, and the chimpanzee also show greater aggressiveness as adults, and more rough-and-tumble play as infants. Rodents too show the same pattern of adult aggressiveness. This aggressive behaviour is well known to depend on the sexual endocrine system. Castration in adult male animals reduces fighting, and injections of testosterone restore the behaviour. Similarly, if the normal processes of sexual differentiation are altered by giving injections of male hormones or by castration soon after birth, the aggressive behaviour is altered in the predicted direction: the former masculinizes females with regard to sexual endocrinology and behaviour, while the latter feminizes males in these respects. In humans, it has been found that females masculinized by exposure prenatally to male hormones show, during childhood, markedly 'tomboyish' behaviour. Much space could be given over to a detailed presentation of the available evidence, but the findings are too clear-cut to make this necessary; aggressiveness is sex-linked in mammals, and is closely associated with sexual hormonal secretions.

The strongly biological basis of aggressive behaviour can be demonstrated by the fact that chemical agents can induce aggression in animals and man, and can also suppress it. LSD-25 and mescaline, as well as alcohol, have been shown to induce aggressive behaviour, and morphine too often has this property. These drugs usually act by favourable combination with experimental or environmental conditions. Such drugs as pyrimidine and imidazoline, however, seem to produce fighting behaviour almost regardless of environmental conditions. Suppression of aggressive behaviour has been reported for some eighty different drugs, although in many cases the action may be due to sedative and other types of action of the drug. However, chlorpromazine and chlordiazepoxide have a 'taming' effect which owes little to sedation. The same is true of such cholinergic blocking drugs as scopolamine. There is little doubt about the fact that chemical agents, whether related to the sex hormones or not, affect aggressive behaviour.

The main reason for mentioning these drugs in the present context, however, is of course the fact that at least some of them are directly related to the sexual hormones. Chlorpromazine, to take but one example, induces pseudo-pregnancy in rats when administered in large doses. In humans, endocrine disturbances have been reported in women, and the urinary excretion of several oestrogens has been found to be diminished by chlorpromazine. In mammals it stimulates the production of milk, indicating clearly the 'feminizing' effect of the drug. It may also produce impotence in men. Thus the same drug produces 'tame', non-aggressive behaviour in humans and animals, and it also stimulates the output of female hormones; it seems reasonable to suppose that in part its 'taming' action depends on the 'feminizing' properties of the drug. It is also of interest to note that chlorpromazine was one of the first phenothiazines to be used in the treatment of schizophrenic psychosis, and its success in this field suggests an action on the P factor we have encountered earlier on in this chapter; P, it will be remembered, was highly correlated with masculinity.

Dominance is another quality which we tend to associate with males, while submission is more characteristic of females; this again holds true not only of humans, but also of many different mammalian species. Social dominance appears to be associated with sex hormone production; much evidence has been produced by experiments with rats in which the animals were injected either with the male sex hormone, testosterone, or with the female sex hormone, oestrogen. The test used in these studies uses a group of, say, eight animals, setting them in pairs to compete for food; winners are considered to be more dominant. The experiment presents some difficulties; it is necessary to control for reinforcement among the competing animals, in the sense that consistent winners or losers tend to establish habit patterns which are difficult to change by hormone injection. One method of doing this is to first establish a firm hierarchy in the group, and then remove the top (dominant) and bottom (submissive) animals. The remaining six animals would then compete against each other, but they would also be set to compete with either the top or bottom animal a number of times depending on the number of wins and losses recorded in their other encounters, in such a way that for each animal the total number of wins and losses would be equal, thus equalizing the effects of reinforcement. When this was done, the effects of oestrogen became very clear—the animals injected with the female hormone consistently occupied the bottom places. The effects of oestrogen are not always observed when precautions of this kind are not taken, but the effects of testosterone are stronger, and do not require such correction for the effects of reinforcements. In any case, the results show clearly that dominance patterns in social behaviour are very dependent on the sex hormones, and that cultural factors, while they may aid and facilitate sex differences, are not by themselves responsible for the observed patterns of dominant and submissive behaviour.

At the other end of the scale from aggressiveness and dominance lie the qualities which psychologists call affiliation and nurturance—the tendencies of one individual to seek the proximity of others and to derive pleasure from doing so, and to help and succour others when in need. Co-operation, too, belongs in this picture, and it will not come as a surprise to most readers to hear that women evince these types of behaviour more frequently than men. Observation of young children shows that this differentiation takes place very early: it is already clearly apparent at the age of three or four. Girls tend to perform a care-taking and protective role; they often play with children younger than themselves, whom they help and assist, while boys tend to play with older children and try to join in their activities. These behaviour patterns are characteristic not only in our society; they have also been found in infants and young children belonging to cultures as varied as India, Okinawa, Mexico, Kenya and the Phillipines. As Corinne Hutt says (1972), 'when we find sex differences which prevail as consistently over time, space and species, as those discussed in this chapter, it seems highly improbable that they are entirely culturally ordained.'

On fearfulness the evidence about sex differences is unequivocal as far as humans are concerned, embracing not only reports on normal people but also evidence of the incidence of depression, anxiety and other neurotic disorders as we have already seen. In animals, evidence is more sparing, but primates would seem to conform to the human pattern, while for rodents there is evidence in the opposite direction—males might be said to be more fearful. In so far as fearfulness is the opposite of aggressiveness, the same evidence already considered points to the implication of sexual endocrines. One interesting study found a direct correlation between lack of testosterone production and non-aggressive, fearful behaviour—although only in younger men; it did not apply to those over thirty years old. Much other evidence is considered in such books as Corinne Hutt's *Males and Females*, which should be consulted for detailed references; she summarizes some of her conclusions as follows:

> The male is physically stronger but less resilient; he is more independent, adventurous and aggressive; he is more ambitious and competitive. . . . The female at the outset possesses those sensory capacities which facilitate interpersonal communion; physically and psychologically she matures more rapidly; she is more nurturant, affiliative, more consistent. . . . For many of these characteristic features there are biological bases. For success, it has been provident to have the more adaptive behaviours under some measure of genetic control. These behaviours are adaptive in terms of the reproductive roles that males and females fulfil: the conformity and consistency of the female makes her a reliable and dependable source of nurture for the infant in its protracted dependency; for more effective communication and socialization a greater emphasis and reliance on linguistic skills and moral propensities has proved valuable; for the exploring and resource-hunting male in turn, a facility for dealing with spatial and conceptual relationships, for reasoning, for divergence in thought and action, has proved equally useful.

Corinne Hutt in this summary already alludes to some of the differences in cognitive abilities to which we must now turn. In overall IQ there is no difference in mean between males and females, although there is some evidence that there are fewer women, compared to men, with very high or very low IQs; these two deficits just about cancel out. They may account for the fact that there are more male mental defectives and also more male geniuses, but there are possible cultural factors to account for these differences, apart from IQ, and the question should not be regarded as settled. It is when we turn to more special ability factors that differences become apparent. At the most primitive level, women have lower touch and pain thresholds than men—in other words, they are more sensitive to touch and pain. Females also hear better than males and their olfactory sense is better developed, but their visual system is less outstanding—men are superior as regards vision. These differences are not due to environmental or cultural influences: they are apparent already in infancy. These differences may be responsible, among other things, for such marked effects as the greater susceptibility of males to visual erotic stimuli. There is some evidence that sensory differentiation of this kind may be dependent on the early organizing action of androgen. Certainly these sex-linked characteristics are not peculiar to the human species, but have also been observed in monkeys and even rats.

When we turn to more complex abilities, we find that men excel in visuo-spatial ability, that is the ability to organize, relate and manipulate visual inputs in their spatial context. Here too animals (in particular, chimpanzees and rats) show the same sex-related pattern, which does not seem affected much by cultural factors. From the point of view of evolution this may be related to the male animal's need to maintain accurate spatial orientation during his foraging, and to detect spatial relationships despite distortions and camouflage. There is evidence both for genetic control of this ability, and also for the suggestion that at least one of the genes controlling it is a recessive carried on the X-chromosome, that is sex-linked. Similarly, there is evidence to show that spatial ability develops under the partial control of the sex hormones. All this makes perfectly good sense in phylogenetic terms.

If men are superior in respect to visuo-spatial ability, women show almost the same degree of superiority with respect to verbal ability. Girls learn to talk earlier, they articulate better, and they acquire a more extensive vocabulary at all ages than do boys. They write and spell better; their grammar is better, and they construct sentences better. The earliest beginnings of this differentiation can be located as early as six months. In other species, particularly in those where the individual's affective state is indicated by characteristic vocalizations, females also show pronounced superiority. All this should not be misinterpreted; females are superior in language usage, or verbal fluency; they are not superior with respect to verbal reasoning, that is the use of intelligence in tasks which are presented verbally. When comprehension and reasoning are taken into account, boys are somewhat superior to girls. Allied to this fact that females are superior with respect to those properties of language which can be learned by rote is the fact that women excel in all rote learning tasks. As Corinne Hutt in *Males and Females* puts it: 'Women are able to hold in their memory store for short periods of time a number of unrelated and personally irrelevant facts, where men are only capable of comparable memory feats if the material is personally relevant or coherent. . . . It seems to be more a sex-dependent facility than a skill readily cultivated in certain occupations.' Corresponding to this stress on rote memory in girls is the fact that in general boys have a more divergent cognitive style, which may suggest greater creativity and originality:

this difference can already be seen in the play of the pre-school child.

Sex-linked personality differences may show themselves in looking at certain occupations. Thus we tend to think of men as outgoing warriors, while artistic pursuits are often considered 'feminine'. If this were true, we would expect men engaging in war-like pursuits to be stable extroverts, while men engaging in artistic pursuits would share the feminine pattern of introverted emotionality. Studies of commandos and paratroopers have shown that almost without exception their personality pattern was that expected in terms of this hypothesis: they all fell into the stable extrovert quadrant. For artistic pursuits, several studies have shown that men who study visual arts tend to fall into the introverted, emotional (melancholic) quadrant. Consider the results of one such study (Götz and Götz, 1973) in which the most original and successful students in an academy for the teaching of painting were selected by their teachers, and then given a personality inventory. It was found that every one of them fell into the 'feminine' quadrant. When a comparison was made of fifty good and fifty poor students, the good students were significantly more introverted and more emotional. These findings are not perhaps surprising, but they indicate the importance of sex-linked personality traits for professional choice, either within or between sexes. Men with a more feminine personality apparently do better in the arts than masculine men; masculine men do better at war-like pursuits. It is not accidental that the nursing profession is largely feminine; it is here that the affiliative and nurturant qualities we have noted before as characterizing women can find their clearest expression.

The evidence here reviewed very briefly leaves no doubt that there are many marked differences between men and women which are due, not to cultural factors (although these may serve to accentuate and even exaggerate already existing biological differences) but to genetic predispositions firmly geared to the endocrine sexual mechanism. These differences are equally apparent in the fields of abilities (both sensori-motor and cognitive) and in that of personality (with particular reference to aggressiveness and nurturance/affiliation). For women to deny these differences, and to seek for a purely masculine role in their efforts to achieve 'equality', is to deny the value of the specific feminine contribution to society; by seeking equality in precisely those fields where nature has seen fit to endow men with greater capacities than women they make certain that women will quite definitely be inferior to men. It is by capitalizing on those areas where nature has seen fit to endow women with greater capacities than men that women can establish their position as equal but different complements to men. They may find that if they cannot beat nature, they might benefit by joining it!

CHAPTER 6

Social consequences

Social policy is—or ought to be—based on two entirely
different kinds of premises. One set of premises is of an
ethical, moral, philosophical kind; this dictates the
direction in which we want to go. The other set of premises is of a factual, empirical, scientific kind; this dictates how we can get to the point we wish to reach—
and it may also tell us whether it is humanly possible to
get there at all. Facts may help us in deciding upon the
means, but they do not furnish us with the ends; morality
tells us about the ends, but leaves us the task of finding
the means. It is only when both sets of premises are kept
in harmony that socially useful and valuable action can
result. Pursuit of the wrong ends, however well chosen
the means, must in the long run be disastrous; pursuit
of the right ends, but through the use of the wrong
means, must end in failure. It is for this reason that social
science is so important; most civilized people are agreed
on the ends—abolition of war, of want, of discrimination
on the grounds of sex, class, creed, or colour—but we
do not have the necessary knowledge of the means
which alone could translate these aspirations into
reality. Only science can furnish us with the needed
knowledge.

It is fashionable to decry science, and blame all our
troubles on modern technology: it is worth while remembering that before the rise of modern science and
technology we were faced with the same problems, often
in a much more acute and difficult form. War, famine,
overpopulation and genocide have always been with us:
science has not created these problems. Science and technology do, however, give us the means of overcoming
them provided we can extend the range of scientific

knowledge to embrace social and psychological factors. We have muddled along for thousands of years without appreciably improving our ability to deal with human problems, or social ones; our educational or penological practices are not noticeably more successful than those of the ancient Greeks, and may be inferior.

Psychologists have spent over a hundred years now in eliciting important facts about human nature; and it seems likely that these facts will have consequences of a social kind. Social policy would certainly be the poorer for not paying attention to these discoveries: educational and other policies, based on false or non-existent knowledge of human nature, are not likely to have the kinds of results that policy makers hope for. This being so, we might have thought that psychologists would be among the first to try and spell out the consequences of their discoveries for education, for penology, and for social policy generally. This has not been so. Psychologists have been only too happy to retreat into their laboratory ivory towers and leave any thinking about possible application of their findings to amateurs and outsiders, usually lacking in specialized knowledge, and often with a sizeable axe to grind. Their motivation, no doubt, was based on humility and modesty, and in truth psychology has much to be modest about. Nevertheless, there are a number of hard facts, and these should be widely known; more, they should form the basis of any thinking about educational and other problems. Only quite recently have psychologists stepped into the arena and begun to think aloud about the application of psychological knowledge to social questions. In this chapter I am going to try and point out some of the implications of the facts surveyed in the previous chapters.

It may be useful to begin with a review of a similar attempt made quite recently by Richard Herrnstein, Professor of Psychology and Chairman of the Department at Harvard University. His article, simply and starkly entitled 'I.Q.', appeared in the September 1971 issue of the *Atlantic Monthly*, and aroused a considerable amount of interest; it also produced the usual bigoted and fanatical reaction among some left-wing students which any mention of facts not in agreement with their prejudices is likely to engender. Having reviewed some of the evidence about the inheritance of intelligence, he goes on to say:

> The specter of Communism was haunting Europe, said Karl Marx and Friedrich Engels in 1848. They could point to the rise of egalitarianism for proof. From Jefferson's 'self-evident truth' of man's equality, to France's *'egalité'* and beyond that to the revolutions that swept Europe as Marx and Engels were proclaiming their *Manifesto*, the central political fact of their times, and ours, has been the rejection of aristocracies and privileged classes, of special rights for 'special' people. The vision of a classless society was the keystone of the Declaration of Independence as well as the *Communist Manifesto*, however different the plan for achieving it.
>
> Against this background, the main significance of intelligence testing is what it says about a society built around human inequalities. The message is so clear that it can be made in the form of a syllogism:
> 1 If differences in mental abilities are inherited, and
> 2 If success requires those abilities, and
> 3 If earnings and prestige depend on success
> 4 Then social standing (which reflects earnings and prestige) will be based to some extent on inherited differences among people.

Herrnstein goes on to deduce five corollaries which aim to apply these premises to future developments:

> (a) As the environment becomes more favorable for the development of intelligence, its heritability will increase, as the preceding section showed.

Regardless of whether this is done by improving educational methods, diet for pregnant women, or whatever, the more advantageous we make the circumstances of life, the more certainly will intellectual differences be inherited. And the greater the heritability, the greater the force of the syllogism.

This is undoubtedly true; even with our existing inequalities of education, nutrition and general upbringing, heredity accounts for twice as much of intellectual development as does environment; with increasing equality the disproportion can only increase.

Herrnstein's second corollary runs as follows:

(b) All modern political credos preach social mobility. The good society should, we believe, allow people to rise (and, by implication if not by frank admission, fall) according to their own efforts. The social barriers of the past—race, religion, nationality, title, inherited wealth—are under continuous assault, at least in principle. The separation of church and state, the graduated income tax, the confiscatory inheritance tax, the laws against discrimination and segregation, the abolition of legal class and caste systems all manifest a desire to accelerate movement on the social ladder. The standard wisdom of our time avows that people should be free of 'unfair' impediments and divested of 'unfair' advantages in all their endeavors. But the syllogism becomes more potent in proportion to the opportunities for social mobility, for it is only when able and energetic individuals can rise and displace the dull and sluggish ones that there can be sorting out of people according to inherited differences. Actual social mobility is blocked by innate human differences after the social and legal impediments are removed.

Here again there is no doubt that Herrnstein is right;

there are limits to social mobility, and these limits will tend to become more and more related to genetic differences between people the more environmental and social barriers are removed. This leads Herrnstein to his third corollary:

> (c) It was noted earlier that there are many bright but poor people even in affluent America. The social ladder is tapered steeply, with far less room at the top than at the bottom. The obvious way to rescue the people at the bottom is to take the taper out of the ladder, which is to say, to increase the aggregate wealth of society so that there is more room at the top. This is, of course, just what has been happening since the Industrial Revolution. But one rarely noted by-product of poverty is that it minimizes the inherited differences between classes by assuring that some bright people will remain at the bottom of the ladder. As the syllogism implies, when a country gains new wealth, it will tend to be gathered in the hands of the natively endowed. In other words, the growth of wealth will recruit for the upper classes precisely those from the lower classes who have the edge in native ability. Whatever else this accomplishes, it will also increase the IQ gap between upper and lower classes, making the social ladder even steeper for those left at the bottom.

The truth of this corollary depends to some extent on the relevance of IQ to social mobility. We have noted already that there is a considerable amount of social mobility, strongly related to IQ; it is quite possible that the social changes discussed by Herrnstein will not have much effect on IQ-determined rises and falls, but rather on personality-determined rises and falls. There will probably be a small change in the direction suggested by him, but it must be doubtful whether this change will be all that marked. Herrnstein goes on to say:

(d) Technological advance changes the marketplace for IQ. Even if every single job lost in automating a factory is replaced by a new job someplace else in a new technology, it is more than likely that some of those put out of the old jobs will not have the IQ for the new ones. Technological unemployment is not just a matter of 'dislocation' or 'retraining' if the jobs created are beyond the native capacity of the newly unemployed. It is much easier to replace men's muscles with machines than to replace their intellects. The computer visionaries believe that their machines will soon be doing our thinking for us too, but in the meantime, backhoes are putting ditchdiggers out of work. And the ones who stay out of work are most likely the ones with the low IQ's. The syllogism implies that in times to come, as technology advances, the tendency to be unemployed may run in the genes of a family about as certainly as bad teeth do now.

Here Herrnstein is definitely beginning to run off the rails in his predictions; as we shall see, there is no real basis for this corollary in the premises. Before turning to a detailed criticism of Herrnstein's thesis, we may, for the sake of completeness, quote his last corollary; it does not add anything to the argument, and is (if we want to be pedantic) not a corollary following from his four premises at all. However that may be, this is what he says:

(e) The syllogism deals manifestly with intelligence. The invention of the intelligence test made it possible to gather the data necessary to back up the three premises. However, there may be other inherited traits that differ among people and contribute to their success in life. Such qualities as temperament, personality, appearance, perhaps even physical strength or endurance, may enter into our strivings for achievement and are to varying degrees inherited. The meritocracy concerns not just inherited intelligence, but all inherited traits affecting success, whether or not we know of their importance or have tests to gauge them.

Herrnstein sums up his argument by saying:

> The syllogism and its corollaries point to a future in which social classes not only continue but become ever more solidly built on inborn differences. As the wealth and complexity of human society grow, there will be precipitated out of the mass of humanity a low-capacity (intellectual and otherwise) residue that may be unable to master the common occupations, cannot compete for success and achievement, and are most likely to be born to parents who have similarly failed. In Aldous Huxley's *Brave New World*, it was malevolent or misguided science that created the 'alphas', 'gammas', and the other distinct types of people. But nature itself is more likely to do the job or something similar, as the less well-known but far more prescient book by Michael Young, *The Rise of the Meritocracy*, has depicted. Young's social-science-fiction tale of the antimeritocratic upheavals of the early twenty-first century is the perfecting setting for his timely neologism, the word 'meritocracy'. The troubles he anticipated, and that the syllogism explains, have already caught the attention of alert social scientists, like Edward Banfield, whose book, *The Unheavenly City* describes the increasingly chronic lower class in America's central cities. While Sunday supplements and popular magazines crank out horror stories about genetic engineering (often with anxious but self-serving testimonials from geneticists), our society may be sorting itself willy-nilly into inherited castes. What is most troubling about this prospect is that the growth of a virtually hereditary meritocracy will arise out of the successful realization of contem-

porary political and social goals. The more we succeed in achieving relatively unimpeded social mobility, adequate wealth, the end of drudgery, and wholesome environment, the more forcefully does the syllogism apply.

Where does Herrnstein go wrong in his somewhat dispiriting evocation of the future of our more and more egalitarian society? The answer would seem to be that he disregards the importance of *regression*, the genetic factor which causes children of very bright and very dull parents to regress towards the mean of the whole population. In his general discussion Herrnstein actually mentions this factor, but he underestimates the size of regression effects very markedly. Assuming that he is right, and that low-IQ workers are likely to form the major part of the pool of unemployed he envisages, it still does not follow that 'the tendency to be unemployed may run in the genes of a family about as certainly as bad teeth do now'. Regression will enable many of the children of the dull unemployed to rise in the social scale; regression will force many of the children of the bright employed to fall into the group of dull unemployed. When Herrnstein talks about 'social classes' he gives the impression (perhaps unwittingly) that these are im-mutable; indeed, he says explicitly that 'our society may be sorting itself willy-nilly into inherited castes'. But that is precisely what cannot happen upon genetic con-siderations; regression makes it quite impossible that castes should be created which will breed true—that is, where the children will have the same IQs as their parents. Within a few generations, the differences in IQ between the children of very bright and very dull parents will have been completely wiped out. Castes can only be created and preserved by social policies insisting on the branding of each child as a member of one caste or another. For anyone wishing to perpetuate class or caste differences, genetics is the real foe; there is no way (except wholesale slaughter) of preserving the *status quo*. Herrnstein's corollaries depend on the picture given in our Figure 7, but as we have seen, this is incorrect; substitute the more correct picture given in Figure 8, and you will see immediately why Herrnstein's fears are unrealistic.

Indeed, we might claim that he is quite wrong in saying that 'The more we succeed in achieving relatively unimpeded social mobility, adequate wealth, the end of drudgery, and wholesome environment, the more forcefully does the syllogism apply'. At present, regression in IQ is to some extent blocked from expressing itself in regression of social class as well; the extent of this failure is premised upon the existence of social barriers to social mobility. When these are removed, then social mobility will be increased and will produce a much greater degree of regression in social class—leading to the very opposite effect to that predicted by Herrnstein. It seems likely that greater regression effects with respect to social class will also lead to a lessening of class conflict, and all its attendant vices. It is when classes are seen as monolithic, persistent, inescapable limitations to change that class loyalties lead to class hostility; this might have been the position in the last century, but it is not the position now. When parents see their children freely change their social class, it will be much more difficult to feel that other classes are different, dangerous forces which engender hostility, opposition, and possibly 'class war'; when one's relatives are freely spread among different classes such absurd attitudes will become of purely historical importance.

But will social classes still exist when egalitarianism has run its course? The data quoted from the orphanage study suggest that differences in IQ will certainly persist, to almost the same extent as now, and according to Herrnstein's syllogism this carries implications regarding social classes. His argument runs as follows:

Are there alternatives short of turning back to social rigidity, poverty, drudgery, and squalor? The first two premises of the syllogism cannot sensibly be challenged, for they are true to some extent now and are likely to become more so in time. The heritability of intelligence will grow as the conditions of life are made more uniformly wholesome, intelligence will play an increasingly important role in occupational success as the menial jobs are taken over by machines. The one even plausible hope is to block the third premise by preventing earnings and prestige from depending upon successful achievement. The socialist dictum, 'From each according to his ability, to each according to his needs', can be seen as a bald denial of the third premise. It states that, whatever a person's achievement, his income (economic, social, and political) is unaffected by his success. Instead, the dictum implies, people will get what they need however they perform, but only so long as they fulfill their abilities. Those in power soon discover that they must insist on a certain level of performance, for what the dictum neglects is that 'ability' is, first of all, widely and innately variable, and secondly, that it expresses itself in labor only for gain. In capitalist countries, the gain is typically in material wealth, but even where the dictum rules (if such places exist), social and political influence or relief from threat would be the reward for accomplishment. Human society has yet to find a working alternative to the carrot and the stick. Meanwhile, the third premise assures the formation of social classes.

Opposition is often expressed to the third and fourth premises of Herrnstein's syllogism on the grounds that it leads to 'elitism' and to the creation of a 'meritocracy'. These are emotion-laden terms, and the emotion aroused may get in the way of a rational discussion of the problems which arise. In particular, those who are opposed to a meritocracy seldom define the alternative state of affairs, which would usually turn out to be what I have called a 'mediocracy'. It may be worth while to consider the two possibilities in relation to particular examples. Suppose that your child is in need of the services of a brain surgeon—perhaps a tumour is developing in the cortex, and only immediate surgery can save the child's life. If you had a choice, would you prefer the operation to be carried out by a surgeon at the top of the 'meritocracy', that is, possessed of the highest ability in his speciality, with the best training, and the greatest experience, or would you take pot luck with just any third-rate medical student? If your plane was getting into trouble, would you prefer a first-class pilot to be at the helm, of proven ability and extensive experience, or would just anyone do? If your life was at stake because you had been accused unjustly of murder, would you prefer to be defended by an outstandingly clever QC, or would a rather dull lawyer be good enough for you? There would be little point in continuing the list. Whenever we need help in our own affairs, we want quite naturally to obtain the services of a person who is best qualified to carry out the job. And note that in these cases the choice is inherently limited to people all of whom are possessed of a reasonable IQ— because no really dull person would be able to pass the stringent examinations which are required for these positions. We could never really be in a position where we would have a free choice between a surgeon with an IQ of 150, and one with an IQ of 50!

Let us generalize the argument a little. Agricultural labourers have a mean IQ of 90, doctors of 125 or thereabouts. While some agricultural labourers would probably have a high enough IQ to become doctors, the majority would never, under any conceivable circumstances, pass the many and difficult tests which society requires (rightly) to be successfully taken by those who are entrusted with our lives in case of illness. On the other hand, most doctors could probably carry out the job of agricultural labourer, particularly if they had never known any other kind of life. Thus genetic factors predetermine some people to be capable of being educated to carry out complex and difficult jobs, while they predetermine other people to be incapable of being so educated. Even if some of the less able could ultimately be made into doctors, they would require a far longer period of training, and they would probably end up as less efficient doctors. This, then, is the principle of meritocracy; that in choosing those whom we wish to become our doctors, our pilots, our lawyers, our scientists, our academics, our professional classes in general, we should pick those who by nature are best fitted to learn the requisite skills quickly, and carry out the jobs most efficiently. Experience shows that intelligence is vitally necessary for the purpose, and the facts show that IQ tests can predict with considerable accuracy just which children have this necessary ability. What is the alternative?

Presumably the principle of mediocracy would be something like that of Buggin's turn—let chance decide who is to become doctor or bus driver, lawyer or dustman, surgeon or miner. Actually nobody has ever spelled out just what mediocracy would do to fill the needed vacancies in society's occupations; this daunting task has always been neglected in favour of sniping at meritocracy. It has never been quite clear what objections there can be to meritocracy, which simply says that the person best qualified for a given job or profession should occupy it, or practice it. What is suggested is that simple merit, rather than parental wealth, or political influence, or other external and irrelevant factors should decide. This seems no more than common sense, and it agrees perfectly with democratic sentiments. Why the vociferous abuse which greets 'elitism' and 'meritocracy'? The merit principle works with much greater force in the fields of sport; what football team would ever thrive on a diet of 'mediocracy'? The amount of selection which is involved in the choice of a particular member of an elite team is almost incredible; by comparison the tests applied to budding medicos amount to very little. But would any spectator at a football match have things any different? Would he prefer to watch Scunthorpe Reserves play Wigan Comprehensive School Juniors, rather than the Cup Final between Sunderland and Leeds? Talent, merit, ability—any system of choice which pays no attention to these largely innate factors must lead to a low and deadly uniform level of mediocrity. Biology sets an absolute barrier to egalitarianism, in life as in sport.

Clearly, if we wish to come to any reasonable conclusions about social policy in regard to education, occupational selection, and vocational guidance, we must tread warily between two opposing myths. The first myth is that it is social class that determines IQ, rather than the other way about. The other myth is that the child's IQ and social class are both the same as that of his parents. Neither myth, as we have seen, is true. Social mobility, and segregation and recombination of genes between them take care of that. A society which would come as near to our egalitarian desires as is biologically attainable would give the greatest scope possible to this social mobility, and would sweep away all those influences which still too frequently stand in the way of the bright working-class child.

For many years IQ tests were instrumental in ensuring a better education for such children than they would otherwise have been able to obtain; the elimination of such tests by left-wing governments is still one of the least intelligible actions taken in the name of advancing the status of the deprived child. As we have seen, IQ tests are not entirely measures of innate ability; the correlation between genotype and phenotype is only .9 or thereabouts. But this correlation is close enough to make the IQ test a genuine instrument of social progress; it depends far, far less than any alternative measure on social factors which would give unfair advantages to the middle-class child. And even the slight correlation with environmental factors that remains could be removed by a suitable weighting of such factors—as was attempted many years ago by Sir Cyril Burt. For all those who wish genuinely to restore to bright working-class children the best opportunities for an education appropriate to their talents, the restoration of IQ tests to their rightful place seems the best, if not the only way.

One interesting alternative has been used in East Germany. In order to give a better chance to working-class children, they issued an edict according to which university places would be allocated to children on a social class basis, with 50% of these places reserved for working-class children. This policy makes it certain that there will be at university a number of working-class youngsters with an ability decidedly inferior to that of a number of middle-class youngsters excluded from the University. It is difficult to see why this procedure should be more just and fair than that customary in England, where there are at university a number of middle-class youngsters with an ability decidedly inferior to that of a number of working-class youngsters excluded from the university. Given that most youngsters of the requisite ability want to go to university, the most equitable method would seem to rule out class (as well as sex, race and other irrelevant considerations) as a qualification, one way or the other, and decide entirely on ability (and perhaps personality, if that could be shown to be relevant to success at university).

Both East Germany and England practise one sort of selection procedure for university entrance: that of requiring the candidate to pass a series of hurdles, and then selecting the ones who do best in this competition. This ensures that only a small percentage of the whole population can enter, but also that of those who enter, only a

8

small percentage fail. The failure rate at British universities has always been much lower than that in French universities, say. The alternative method of selection is practised in France, and most of all in the USA. At first sight this method seems not to imply selection at all; anyone who passes high school successfully (and this is a very low qualification indeed) is legally entitled, in most American States, to go to his or her State university (there are different rules in different States). But of course the reckoning has to come some time; an impossibly large proportion of the students who so enthusiastically throng the campus at the beginning flunk out after one or two years. Selection has simply been postponed, with the result that thousands of youngsters lose valuable time, when they can least afford it, and are labelled, by others and by themselves, as intellectual failures. This is a high price to pay for the neglect of proper selection procedures, and it probably accounts for at least some of the unrest that was so apparent particularly in American and French universities.

A new method of selection has recently made its appearance in the USA, modelled after the 'class selection method' prevalent in East Germany, and described above. That is the so-called 'quota' system, according to which placement in school, acceptance at university, staff membership of school and university teaching groups, and even election to political party conventions is dependent on the observation of 'proportional representation' of minority groups. This movement began with the observation that black children were over-represented in classrooms set aside for remedial education; while 9·5% of the school age children in California are black, for instance, 27·5% of the children in the State's remedial classes are black. This led to legal action on the part of blacks ruling illegal the practice of placing children in such classes partly on the basis of IQ tests; in fact, in some States the administering of IQ tests to black pupils has been declared illegal. The next step was taken by some school districts, such as that of Berkeley in California, where quotas were legally enforced in state schools: according to these rules, the proportion of blacks in remedial and in specially advanced classes had to be the same as the proportion of blacks in the population. In England this stage has not yet been reached, although there have been many complaints by coloured communities that the proportion of coloured pupils in ESN (educationally subnormal) classes is excessive.

The effects of such a quota system, although this has not been in action for very long, have already become apparent. Black children in need of special remedial education are deprived of this, and are sent to classrooms where the teaching is above their heads, with the obvious result that they become difficult to teach, and run riot, making it difficult or impossible for normal teaching of the other children (white and black) to proceed normally. Black children who would have managed to get along quite well in ordinary classes are suddenly forced to compete with exceptionally bright children in educationally advanced classes, and again get lost and make normal teaching impossible. Both white and black children lose in this game: no one gains.

One interesting sequel to this chain of events has been that in Berkeley, which was one of the first cities in the USA to desegregate its schools completely, the black community insisted on having a completely segregated all-black unit, called *Black House*. The student body consists largely of those blacks who could not compete in the integrated Berkeley High School. This school is largely funded by the Ford Foundation, which, interestingly enough, was not willing to fund research on the effects of school integration in Berkeley. It is so easy to believe that integration (which I believe to be a desirable thing) solves all the problems of coexistence; instead, it poses many problems which are quite intractable and require large-scale research. Integration by itself is not enough, and without research into the precise consequences of integration we will never even know exactly what these problems are.

At universities, the Nixon administration has inaugurated a programme of 'affirmative action', which means in effect that Federal money (which is vital to the survival of the great majority of universities) is only paid provided that they establish 'goals' (a euphemistic term equivalent to 'quotas') for the hiring of minority group members to staff positions, and actually demonstrate progression towards the achievement of these 'goals'. Such an 'affirmative action compliance program' includes an analysis of 'deficiencies' in the utilization of racial minorities and women, as well as a time-table for the correction of these deficiencies. Among other requirements, the universities (and other corporations: the law applies to all corporations having more than fifty employees and a Federal contract of at least fifty thousand dollars) must advertise to their employees their progress in eliminating the deficiencies, and must give 'active support to local and national community action programs. . . .' The attempts to make a distinction between 'quotas' (which are illegal in the USA under title VII of the Civil Rights Act of 1964) and 'goals', which are set up under coercion by the employing agency itself, are difficult to follow but not perhaps too relevant to the subject matter of this book. Suffice it to say that the American Government is in fact enforcing a system of employment on the universities under which they are required, under pain of bankruptcy, to employ members of minority groups in spite of the fact that a better qualified member of a non-minority group is applying for the job. (Women, for some inscrutable reason, are listed as one of the 'Minority' groups!)

There are two sets of objections to these practices. The first is a simple liberal belief that it is the quality of the individual, his abilities, his personality, his experience in life, and in the particular subject matter which is relevant to the job in question, which should decide his employment or promotion: that factors such as race, sex and religion are irrelevant, and should not be given any weight in coming to such decisions. Quotas were considered undesirable when they were used against minority groups; they do not become desirable when they are used against majority groups. Positive discrimination, so called, is still discrimination against somebody; one man's positive discrimination is another man's negative discrimination. Furthermore, who shall define a minority? In current American practice, blacks, chicanos, American Indians, Puerto Ricans and women are considered minorities, but Italians, Armenians, Arabs, Jews, Irish, Mormons, labour union members and the aged are not. Why are some minorities more minor than others? The quality of justice is not too apparent in these time-serving ordinances.

The second set of objections is simply on the grounds of biological inequality. Some people are better suited to a given position than others; some children require remedial education, others are able to benefit from advanced tuition. It is not possible to alter these facts by legal requirement. You can take the child requiring remedial education, and put him into an ordinary classroom, but you cannot make him benefit from the sort of teaching which is geared to the average child. You can appoint a minority member to a professorial chair which, without the benefit of enforced 'affirmative action' he would not have been considered for on grounds of merit, but you cannot make him equal in ability and performance to the better qualified person he displaces. There can be no question but that these practices will lower drastically the quality of education available to children and university students alike. This is a serious consequence of disregarding the facts of biological inequality. Injustice coupled with inefficiency is a bad combination; this is but one example of the consequences to which disregard of factual evidence may lead.

It is sometimes denied that 'affirmative action' is in fact equivalent to a quota system, and it is asserted that it merely requires such action when applicants are equal in merit. I have received unequivocal testimony that in many cases this simply is not so, and that University Departments are under great pressure to take on staff belonging to minority groups to the exclusion of much better qualified members of what I suppose must be called majority groups. It is not suggested that this is universal, or that all minority staff so taken on are inferior to alternative 'majority group' applicants. However, great injustices are undoubtedly committed in the name of 'affirmative action', and the effects can already be seen.

Why, one must ask, have the ears of so many people in responsible positions been closed to the rational arguments brought forward against the exclusively environmentalist position with regard to intelligence, personality, mental disorder, criminality, and the various other areas discussed? One obvious reason is the apparent misconception that only the environmental side offers us a chance to correct what we regard as social abuses; nothing, so it would appear, can be done about the genetic side—except perhaps 'genetic engineering' which is clearly not practicable at the moment, and which is in any case regarded with the deepest suspicion by most people. This absolute confrontation between heredity and environment is a false antithesis: it depends for its plausibility on an erroneous conception of the term, environment. It is vital for a proper understanding of the issues involved that this misconception be clearly understood; unfortunately there is very little in the literature which aids us in clearing up this misunderstanding. We may perhaps begin by referring to an example of environmental manipulation of clearly genetic causes, namely the case of diabetes.

This is an inherited metabolic disorder, usually fatal; however once an understanding was acquired of precisely what the nature of the metabolic defect was, a cure became possible, and now insulin keeps sufferers of diabetes alive and healthy. (Cure is perhaps too strong a term, as constant administration of insulin is necessary; nevertheless, from the point of view of the patient, he is enabled to live a full and essentially healthy life.) One might also argue about the proposition that diabetes is inherited; it would be more accurate to say that the *predisposition* to diabetes is inherited; whether or not the disease develops depends largely upon the way the body reacts to stress. In any case, the outstanding feature of diabetes is an increase in blood sugar, so that eventually much of this sugar spills over into the urine. Insulin is given to diabetics because the pancreatic hormone diminishes the blood sugar and improves its utilization as fuel for the tissues. Recognition of the hereditary nature of the disease opened the way to a study of the nature of the inherited defect, and this in turn made possible intervention and finally 'cure'.

Treatment of the psychotic disorders (schizophrenia, manic-depressive illness) seems to be going a similar course, although the causal chains are less clearly understood than in the case of diabetes. The tremendous success of drug treatment (phenothiazine derivatives; Rauwolfia alkaloids; tricyclic compounds; monoamine oxidase inhibitors; etcetera) has been dependent on the recognition of a strong hereditary basis for these disorders, and consequently a biological orientation towards their treatment. Environmental theories, such as those of the psychoanalysts, have produced no answers whatsoever for the tremendous problem produced by these disorders; they have only served to delay proper scientific investigation by many years, and have completely failed even to suggest specific environmental influences which might work with a hereditary predisposition to bring on active psychosis. But the recognition that hereditary influences are at work does not mean that we must give up hope of finding a cure. It is the beginning of wisdom. Once the biological nature of the complaint is recognized, we can investigate it in sufficient detail to specify what is wrong: we can then set about putting things right.

An even simpler example is provided by one form of asthma which is known to have a genetic basis. At one time psychoanalysts assumed that asthmatic attacks in the children afflicted were produced by their mothers who had emotionally mishandled the upbringing of the children in question; as proof they cited the fact that the children often got better when away on holiday, or with relatives. We now know that what is responsible is the house mite, a microscopic organism which flourishes in the dust which accumulates in some mattresses; inhaling these mites, or even small bits of dead ones, brings on asthmatic attacks in children genetically predisposed in that direction. Going on holiday, or to relatives, takes the children away from the mattresses and the mites, and allows them to recover from the attacks.

In a similar vein, hay fever has a strong genetic predisposition at its base; the sufferer is liable to develop strong reactions to breathing in pollen. Every summer quite a colony of hay fever sufferers congregate in Aberdeen, where the wind usually comes from the east, over a large ocean surface, and sheds any pollen it may have acquired on the way. Once the genetic basis of the disorder has been recognized, we can alter the environment (clean the mattresses, live in Aberdeen) in such a way that the individual predisposition to react in undesirable ways is not activated. Allergies provide another example: again we are dealing with genetic predispositions to react strongly to specific types of stimuli, and a precise diagnosis can be made by testing the reaction of the skin to a variety of different materials. Once the guilty substance has been identified, a programme of desensitization can be undertaken, and the sufferer cured.

An example of how this sort of conception can be used

in working out methods of dealing with social problems is given in my book on *Crime and Personality*. We have already noted that criminals tend to come from the unstable, extrovert quadrant; I have argued that such people are characterized by low cortical arousal, and that low cortical arousal prevents them from forming conditioned reflexes readily. Now for most people honesty and other social virtues are probably encouraged and maintained by some form of 'conscience' which, according to theory, is itself the product of a long history of conditioning—through parental action, through peer mediation, through teachers' influence, etcetera. Extroverts, on this account, are handicapped in acquiring a well-developed conscience by their innately poor conditionability; hence they are much more likely to fall into temptation, and carry out criminal, or at least anti-social, acts. Neurotics, on the other hand, are introverted for the most part, and hence have a cortex which is in a steady state of high arousal. They condition only too well, and hence acquire neurotic symptoms, which are conceived of as simple conditioned emotional responses, very easily. This theory, which is here presented only very briefly, and in a grossly oversimplified form, would also account for the strong genetic component in both neurosis and crime. (I have argued the case in detail in *Crime and personality*.) Does this mean that nothing can be done to prevent or cure neurosis and crime?

The answer surely must be that if our theory of just how heredity influences our behaviour in the direction of developing criminal or neurotic behaviour is correct (and for the purpose of this discussion I shall assume that this is so, although of course the case is far from proven) then it also suggests ways and means of preventing or curing both disorders. Taking criminality first, we note that extroverts condition less quickly and less strongly: this, according to theory, is an innate factor which we cannot touch directly. But extroverts *do* condition, if slowly; we can produce the same degree of conditioning in an extrovert by submitting him to a course of 300 pairings of the conditioned and the unconditioned stimulus, as for example, the bell and the food in Pavlov's famous experiment, as we could in an introvert by submitting him to 50 pairings, say. In other words, extroverts need a firmer, more consistent type of upbringing in order to produce the same effect that in the introvert would be produced by a more lenient and possibly less consistent kind of upbringing.

So much for prophylaxis; how about treatment? Once the criminal is convicted and imprisoned, we clearly should try to do for him what his parents, teachers and peers have clearly failed to do, that is inculcate in him some form of 'conscience'. This can be done by a detailed, consistent and well-planned programme of conditioning, such as I have described in *Psychology is about People* (1972). Note that such a programme would not have in it any of the crudity and cruelty which characterized the purely imaginary methods used in the film, *The Clockwork Orange*. Not only would such methods be abhorrent, but they would also be useless. Thus recognition of the precise nature of the inherited weakness of the extroverted person leads us to the design of environmental methods of manipulation more likely to be efficacious than those used in the past, without benefit of such knowledge.

What in fact would the psychologist proceeding along these lines do? *The Clockwork Orange*, true to the sensationalizing and pornographic tradition of some modern film-makers, immediately brings to mind paralysing electric shocks which reduce the sufferer to a gibbering vegetable. This sort of nonsense recalls my first venture into public advocacy, when I discussed the 'bell-and-blanket' method of curing enuresis. (What is done is to make the child sleep on a padded wire-mesh mat which is connected to a battery and a bell. When the child wets the bed, the first few drops make an electric contact, the bell rings, and the child wakes up, stops urinating re-flexly, and goes to the toilet. Gradually, this waking-up reaction becomes conditioned to the feel of the distended bladder, and the enuresis is cured.) Immediately a cry of horror went up from psychotherapists who had tried for many years to treat enuresis by Freudian methods (without success) and who complained publicly about the cruelty of giving children electric shocks to their private parts! In this way a humane, quick-working and entirely safe and non-painful method of treatment is transformed into a malicious, cruel, pornographic type of torture which no right-thinking parent would con-template for one moment. Much the same seems to have happened to my suggestion that some form of 'con-ditioning' should be applied to convicts otherwise apparently impossible to rehabilitate.

What is intended, rather, is some form of 'token economy', such as has been proving extremely successful with deteriorated schizophrenics, bringing them back to useful, enjoyable living, and which has been applied with considerable success to criminals (usually adoles-cent ones so far) in the United States. This method, first introduced about a hundred years ago by Alexander Maconochie into the large convict colony at Norfolk Island, and immediately successful there, in spite of almost insuperable obstacles, rests on the belief that criminals have failed to learn the 'contingencies' of ordinary life. The term 'contingencies' has a technical meaning here: roughly, it refers to the sequence of con-ditioned and unconditioned stimuli and responses which go to make up the paradigm of the conditioned re-sponse. Whether through genetic defects (as I believe to be true in many cases) or through a particular history of faulty reinforcement (as Skinner would maintain) or probably both in varying proportions, criminals have failed to acquire a proper sense of the contingencies involved in ordinary living; in particular, the sequences *work → reward* and *crime → punishment* (which contain the essence of 'conscience' and socialized living) have not been properly conditioned in these people. (Note that this statement does not imply any blame; neither the criminal's heredity nor his 'history of reinforcements' are under his control in any meaningful sense. The notion of blame (and free will) is essential to any doctrine of retribution, and both Skinner and I would view such a doctrine as barbaric and also counter-productive.) What is needed, consequently, is an enriched programme in which the partly missing 'history of reinforcement' is recapitulated in such a form as to give rise to the optimal degree of conditioning.

In detail, this means that the convict is given a chance to obtain tokens by work and strict adherence to certain rules (non-violent conduct; calm, quiet behaviour; co-operation; etcetera), and that he is able to exchange these tokens against a variety of self-chosen 'goodies', ranging from periods of television viewing to cigarettes, or from sweets to going out on parole for limited periods. Maconochie originally suggested that each convict could 'work his passage' in this way. He would be freed when he had accumulated sufficient tokens. (The Home Office, horrified, made it impossible for him to carry out this plan.) Of particular importance in Maconochie's project was the allocation of tokens for work well done; the notion of training convicts in decent work methods, and in the *work* → *reward* contingency, was central to his thinking. Skinner and his followers have taken over this whole system and rephrased it in modern psychological jargon (which I have partly followed in the above brief description); it certainly fits in very well with present-day psychological thinking. Does it work?

We must be careful not to be carried away by the enthusiasm which has been shown by some of the psychologists and criminologists who have pioneered such systems, and who have observed its effects at first hand. In view of the short time since modern methods of 'token economy' have been applied in prisons and other correctional institutions, any proper long-term follow-up is of course out of the question. What we do know is that within the prison or correctional institution the system makes a profound and immediate change in the behaviour and outlook of the prisoners; behaviour and work record improve to an extent which many observers would have thought impossible. Short-term follow-up records suggest that even after leaving the correctional system, and the token economy, those who have experienced it have a much better record than control groups not exposed to it. Whether this difference would extend to the end of a twenty- or thirty-year follow-up it is of course impossible to say; clearly work on this should be set up immediately. Whether token economies would work as well with seasoned criminals as with 'beginners' is not known, and may perhaps be doubted; obviously the attempt must be made, and only time will tell to what degree it may be successful. But perhaps the knowledge that at long last something positive can be done with the young offender, who at present only too frequently turns into the old lag through his experiences in prison or Borstal, is sufficient incitement to continue work along these lines, to try to improve our methods, and to set up larger and better units for rehabilitation.

It is noticeable that all this work has been done in the USA, although the original ideas came from Britain, and although the first attempt to put into practice notions such as those underlying the 'token economy' was made by an official of Her Majesty's Government. One realizes of course the difficulties of such research, involving as it does finding work for the young offender to do which is reasonable in itself, which appeals to him as such, and which furnishes him with a proper training which will prove useful in later life—as well as setting up work habits through a process of conditioning. One realizes the difficulties in finding psychologists with any degree of experience in work of this type (which sounds easier than it is in actual practice), and with a proper background in learning theory (many of them are still stuffed with the moribund theories of 'dynamic' psychology). But when all is said and done, our common humanity, as well as strict principles of accountancy, would push us towards the attempt. The cost of 'curing' or rehabilitating a young offender is minimal as compared with the cost of coping with his turning into an old lag; on financial grounds alone the experiment would be well worth doing. But much more important (to me at least) is the opportunity it would give us of saving a youngster from a life which in the long run would offer him nothing but humiliation, unhappiness, and despair. (The romantic picture of crime painted by our television serials can hardly be said to mirror reality.)

I have stressed token economies in this account, although there are other ways of using the principles of conditioning; none of them would follow the lines of *The Clockwork Orange*. It is not only that the methods there depicted would offend our ethical code, and be unacceptable to civilized people everywhere; that much, I imagine, is common ground. But in addition the method is unacceptable on simple psychological grounds; aversion therapy would not work with people who were forced to submit, rather than choosing freely to co-operate with the psychologist. I know of course the difficulties of defining properly what is meant by 'free choice' in this connection, but in practice there is seldom any real question. It is not difficult to see whether a 'patient' comes with the genuine intention of desiring change, or whether he is coming under threat of prison or court action. Even the methods of the token economy should not be forced on unwilling convicts; they ought to be given a free choice of whether they wished to enter the programme, or whether they wished to remain as before. From the American experience there seems to be little doubt that the vast majority would make the choice that gave them a chance of rehabilitation, and there seems to be equally little doubt that they would not regret this choice.

While many, if not most, criminals would be covered by the theory and the methods of treatment here outlined, it is clear that there are other groups of criminals in whom the reasons for their criminal conduct arise from quite other causes, and who require quite different treatment. (Murderers, for instance, tend to be introverted—at least in Europe. Professional killers may be different.) Individual differences are just as pronounced within the criminal group as outside it, and methods of treatment must always be geared to a recognition of the inequality of man—even though this man might be in prison.

Professor I. Sarason, working in Seattle (his research is as yet unpublished) has argued that many juvenile delinquents misbehaved because they had never learned social skills which might enable them to deal with difficult situations at school, at work, and in the home. He has suggested a programme of 'modelling', making use of the important principle of imitation. Juvenile delinquents in groups would be shown a scene such as might happen to them all, with a 'model' showing them how such a scene should be handled. For instance, the scene might depict a boy going for an employment interview; he is kept waiting, he is treated discourteously, there is a mix-up about his qualifications, and some other difficulties arise. How should he deal with these problems? A dull, introverted boy would find great difficulties in coping in such a situation, and might thus throw away a chance of getting a worthwhile job: training in dealing with the problems thrown up by life might give him a better chance.

There are many problems, such as dealing with unfair treatment from a superior or teacher, which arise in the course of one's school and work life, but which few people know how to resolve. This art is particularly important for an adolescent who already has a conviction;

few people will exert themselves on his behalf. Resisting peer pressure to do something which might lead to trouble is another skill which requires teaching. Schools are not concerned with teaching such useful things! We teach children knowledge; we do not teach them adaptive behaviour.

Sarason followed up the young delinquents so taught for three years, and compared their fate with that of a control group not so taught; there was a 50% improvement in the behaviour after release of the group that had received the 'modelling' treatment as compared with the controls. No such improvement has ever been reported after Borstal treatment, or any other official method of dealing with criminality! The interesting point is that Sarason reports that the adolescents for whom this treatment was most effective were anxious, inadequate, neurotic boys, that is the introverted as opposed to the extroverted kind of boy. Psychopathic, extroverted boys did know how to handle social situations from the beginning, and did not require to be taught! However, they also did not benefit from the treatment: for them a strict method of token economy was found vastly superior.

In other words, in order to rehabilitate criminals we must first of all have a theory concerning the causes of their behaviour. We must then apply this theory without failing to take into account individual differences, and build up methods of treatment which, while entirely humane and even pleasant, can also be shown to be effective. (The delinquents were not obliged to attend the 'modelling' demonstrations; they came because they found them interesting and useful. Similarly, delinquents usually like 'token economies', and greatly prefer them to the usual run of treatments. The boys in the 'modelling' treatment experiment were routinely given psychotherapy in the institution where the experiment was conducted. They disliked it intensely, and had to be forced to take part.) It is interesting to speculate why no similar work is proceeding in this country, and why the results of experiments done in the USA are not being applied here. When we consider the vast sum which society could save by rehabilitating even a few of the adolescents who grow up into adult offenders, one may wonder whether we are really serious in saying that we wish to rehabilitate criminals. Perhaps we are in reality only interested in punishing them.

Similar arguments can be put forward in relation to neurosis. Thus if neurotic symptoms are not symptomatic of any underlying 'complex', as Freud thought, but are merely conditioned emotional responses, then clearly simple extinction methods, modelled on laboratory procedures used with experimental conditioned responses, are likely to cure such patients of their troubles. Much evidence is now available to show that 'behaviour therapy', which is the term used to characterize all the methods worked out in this way, is not only more effective than older methods of psychotherapy, but also much quicker. A recognition of the genetic basis of neurotic disorders, together with a clear hypothesis as to precisely what it is that is inherited, and causally related to the neurotic dysfunction, can lead us to methods of behaviour modification clearly superior to previously available types of therapy. Thus the recognition of genetic causes for certain types of behaviour does not lead to therapeutic nihilism; quite the opposite is true. It is the disregard of such genetic causes, coupled with an undue prominence given to environmental causes, that has been responsible for the failure of psychotherapeutic methods of treatment over the past fifty years.

The last few paragraphs may have given a wrong impression, because of their brevity, of the essentially interactionistic nature of the hypotheses employed. It is not suggested that a person's conditionability is the only, or even the main, factor which produces criminal or neurotic behaviour. Obviously environmental conditions are of considerable importance; conditionability, in order to produce actual conditioning, requires the bringing together of conditioned and unconditioned stimulus in the presence of the organism; this bringing together is of course an environmental event. Stress has been put on the genetic aspect of the situation purely because this has usually been neglected in discussions of neurosis and crime during the past fifty years; the intention is to reverse the swing of the pendulum, but not to bring it back to a purely genetic position. We must learn to recognize the importance of interaction in regard to all aspects of behaviour; it clearly will not do to slight either the importance of heredity or that of environment.

One difficulty obscures the argument, and it is important to recognize the existence of this difficulty. When people talk about the 'environment', they often have totally different and often contradictory things in mind. They often mix up the existing environment and some other, possible environment which we could construct on the basis of existing knowledge. They also mix up both of these environments with another one which possesses certain clearly defined end-states, but which we do not know how to construct. And they often endow such possible but non-existent environments with properties which on the basis of existing knowledge they are in fact very unlikely to possess. Worst of all, discussions of the environment and its influence seldom refer to specified and clearly defined entities; the environment usually emerges as a completely vague and unidentified mish-mash of all sorts of influences, many of which remain completely ambiguous, including for good measure a 'factor X' which is invoked when all more clearly defined factors have been shown to be ineffective. One example has already been given to show how erroneous presuppositions about the strength of environmental determination of IQ has led egalitarians to assume that a society which had succeeded in providing environmental conditions of such equality for its members as obtain in an orphanage would lead to a marked shrinkage of the IQ variance, whereas in actual fact any such shrinkage can be shown to be very modest indeed. Let us now consider a rather more esoteric example.

In 1963, Professor O. S. Heyns published a monograph on 'Abdominal Decompression' in Johannesburg, in which he outlined his method of maternal antenatal decompression. In its present form, this method comprises an air-tight plastic suit enveloping the patient from the armpits to the feet, with a rigid dome-shaped spacer over the abdomen. A rubber tube connects this suit to a suction apparatus by means of which, under the direct control of the patient, air can be removed from the suit, thus reducing the pressure over the abdomen by between one and two pounds per square inch. Heyns explains that the effect is to change the shape of the uterus from ellipsoidal to more or less spherical, with concomitant physiological changes, the most important of which from our point of view is an improved blood supply to the foetus and placenta. This is particularly beneficial in cases of toxaemia of pregnancy and placental insufficiency. The improvement of oxygen supply to the foetus, especially to its brain, is the basis of the claim that decompression, as a regular treatment from the twenty-eighth week of pregnancy, actually improves the intelligence of the child. Heyns claimed to have shown a clear superiority of children born after this treatment. Using the Gesell method of estimating the mental qualities of 324 infants under 24 months of age, and using a proportion score similar to the IQ which he calls an F score, Heyns obtained means of 107 for two control groups not receiving the treatment, and scores of 126, 131 and 134 for three groups receiving respectively 14, 15, and 58 decompression sessions. He also reported very significant correlations between F scores and numbers of decompression treatments for various different groups of mothers.

Unfortunately there are many sources of error in this study. Heyns used only volunteers for his decompression group, and subsequent studies have shown that such volunteers tend to be highly intelligent, middle-class mothers; this would tend to result in brighter-than-average babies. The scale used relies to some extent on parental report, and this may have been biased by expectations. Last, intelligence test scores at the age of two are very poor predictors of adult intelligence, the behaviour observed is almost purely physical, and not really prognostic of mental and linguistic development. When R. Liddicoat repeated the study, he obviated these criticisms as far as possible, and obtained the results shown in Table 13. The test given to the mothers was the Wechsler-Bellevue; that given to the offspring at the age of three years was the Merrill-Palmer test, and that given to the children at the younger ages was the South African Child Development Scale, which does not rely for any of its scores on parental report. It will be seen that there is no evidence here for any improvement in IQ due to the decompression treatment. Granted that this new method is no cure for low IQ, it is interesting to consider for a moment what the position would be with respect to the nature-nurture controversy if Heyns' claims had been substantiated.

Clearly, if the treatment were given universally, and if it produced an improvement of 30% in each child regardless of race, or parental IQ, then the population mean would be shifted upwards by something like 30 points. This would not reduce the variance, however, and the general conclusion that 80% of the variance was contributed by genetic factors would still stand. But let us assume that the treatment were only given to black mothers, in the hope that this would equalize the differences in IQ between American blacks and American whites. The effect would be rather startling, improving the mean IQ of the blacks well above that of the whites. Now the proportion of the variance attributable to environment within each race would still remain 20%, but between races it would almost completely eclipse that due to heredity. By varying the proportion of blacks

Sample	Assessment	Decompression group		Controls	
		Mean	S.D.	Mean	S.D.
Mothers	W.B. IQ	96·0	Not given	96·2	Not given
Offspring 1 month	Dev. test scores	32·888	7·026	31·779	7·122
Offspring 4 months	Dev. test scores	96·073	14·262	94·744	14·293
Offspring 9 months	Dev. test scores	50·710	10·060	49·518	10·823
Offspring 3 years	M-P IQ	103·904	11·890	103·155	13·023

Table 13 *Liddicoat's Study (1968)*

and whites 'decompressed', and by varying the number of decompression treatments given, almost any result could be produced regarding the relative contributions of genetic and environmental factors to the growing child's intelligence. But none of this would invalidate the statements made in this book regarding the *present* importance of heredity. There exists an infinitude of environments; the particular environment existing at the moment that the studies here reported were done might be called E_1; for E_1 the proportion of the IQ variance due to genetic causes is 80%. The environment incorporating some new development like decompression, assuming this actually to affect IQ would be E_2; the proportion of the IQ variance due to genetic causes in E_2 could be anything from 0% to 100%. But of course this is looking into a future where some way has been found of affecting IQ, and it is unlikely that such a method will be found which is not based on the recognition of the strong genetic basis for IQ, and a better understanding of just what this basis is. The work on evoked potentials is a first beginning to such a better understanding.

We have seen that decompression is not the answer. Is there any hint as to what the answer might be? There exists now a good deal of evidence to suggest that glutamic acid may be able to raise the IQ of dull children (and rats!) to a very significant extent; the drug does not seem to affect average or bright children (or rats). It is unlikely that this effect is produced by alleviating some pre-existing dietary deficiency in glutamate. It seems much more likely that the improvement in learning ability may be due to the facilitatory effect of glutamic acid upon certain metabolic processes underlying neural activity. It has for instance been shown that glutamic acid is important in the synthesis of acetylcholine, a chemical substance necessary for the production of various electrical changes occurring during neural transmission. The rate of acetylcholine formation can be increased four or five times by adding glutamic acid to dialysed extracts of rat brain. It has also been shown that the concentration of glutamic acid in the brain is disproportionately high as compared with the concentration of other amino acids or with its concentration in other body tissues. Furthermore, of all the amino acids, glutamic acid alone is capable of serving as the respiratory substrate of the brain in lieu of glucose; this further points to the involvement of glutamic acid in neural function. Last, it has been found that the acid exerts its main action on the cerebral cortex, lowering the threshold of excitability. All this points clearly to the importance of glutamic acid in cerebral metabolism, and links up quite promisingly with the work reported on evoked potentials.

Why does it only work with sub-normal organisms? It seems likely that the cerebral metabolism of such organisms is defective in some way while that of average rats and humans is normal, so that glutamic acid could help the defective cerebral metabolism of the dull organisms, while having no particular effect on the normal ones. There certainly exists a relationship between cerebral metabolism and mental functioning. When tissue preparations from the brains of mentally retarded persons are studied, it can be shown that these tissues are incapable of utilizing normal amounts of oxygen and carbohydrates. In cases of mongolian idiocy and phenylpyruvic oligophrenia, for instance, the brain removed much less than the normal amounts of oxygen and glucose from a given volume of blood passing through it. Plainly, the cerebral metabolism of these mentally retarded patients was defective.

We can now envisage a world in which E_x is such that all children with low IQs would receive glutamic acid in sufficient doses to have maximum effect, and for the purpose of the illustration we may imagine that the effect is inversely proportional to the child's IQ, provided it is below 100. In other words, for average and above-average bright children, there is no effect; for below-average children the effect is greatest for the dullest, and least for the relatively bright. In such a world, the mean IQ of the population would be raised and the variance would be smaller; this is precisely the kind of effect that egalitarians are searching for. Now it is not claimed that glutamic acid will achieve all this; at the moment there simply is not enough evidence to make any far-reaching claims for it. Possibly it only works with mental defectives; it is certainly with such children that the most positive results have been reported. Possibly the metabolic defects which glutamic acid may cure are associated with special genes which are not identical with those which cause the normal distribution of intelligence. There are all sorts of questions which remain for proper research to answer. But one question suggests itself above all others. Why, since the effects of glutamic acid on IQ have been known for over twenty-five years, has there not been a determined research effort to answer all the factual questions which arise? There have been over a thousand different 'Headstart' type educational investigations carried out, with practically no effect on the IQs or academic achievements of the hundreds of thousands of children concerned. Why was not some of the money involved, which amounts to hundreds of millions of dollars, spent on the quite straightforward type of research that would be needed to find out once and for all precisely what are the effects of glutamic acid on the intelligence of dull children? It seems just one form of the environmentalist preconception that makes governments and bodies that give research grants open their purses for demonstrably useless types of educational investigations, doomed to ignominious failure from the start, and to pass by biological studies which hold out the promise of success. Education is 'in', drugs are 'out'; that seems to be the message.

Yet one might argue that any egalitarian worth his

salt would jump at the opportunity offered by the existence of drugs such as glutamic acid. What should we be
doing in order to discover its potentialities? We should
try out different doses, for different lengths of time, on
selected children of different IQ levels, and different
ages, measuring any possible effects on IQ, on special
abilities, on school performance, and particularly on the
speed, accuracy and persistence components of intelligence. We should get our physiologists busy with studies
of the precise effects of the drug on the cortex, making
use of such techniques as evoked potentials, as well as
the more traditional ones. We should get our bio-
chemists busy with studies of the precise basis for the
effects observed, and get them to improve the action of
the drug by making slight changes in its formula. It is
never wise to predict what might happen once research
into a novel area is carried out, but I have no hesitation
in saying that the knowledge so gained would be of much
greater social use and value than that acquired by the
thousand and one 'Headstart' type projects.

This abhorrence of the use of drugs is equally obvious,
and in some ways surprising, in relation to criminality.
It is theoretically predictable, and observationally true,
that amphetamine and other stimulant drugs serve to
render obstreperous, hostile and aggressive criminals
co-operative, modifiable, and reasonable. At the same
time, it has been demonstrated experimentally that
these drugs make people more 'introverted', and thus
increase their conditionability, that is the responsiveness
to experiences designed to rehabilitate them is greater.
In spite of these well-authenticated facts, no research is
at the moment being done in prisons or Borstals to make
use of these important aids to rehabilitation. Similarly,
it is well known that phenothiazines and other anti-
psychotic drugs have a tremendous effect on psychotic
disorders; why are these not being studied as possible
aids in dealing with the prisoner with a high P score?
There is a strong possibility that in this way we might

be able to rehabilitate just the kind of prisoner whom everyone has given up for lost and damned. By refusing to face our biological descent and nature, we are still refusing to accept Darwin's evolutionary doctrine; is it not time we gave up the unequal struggle and reconciled ourselves to reality? If drugs can interact with our hereditary nature to produce desirable consequences in conduct, should we let prejudice stand in our way?

However that may be, many people will be somewhat incredulous when faced with the apparent fact, which emerges from our discussion of 'performance contracting' and the various 'Headstart' programmes, that psychology apparently has no real contribution to make to the improvement of education, particularly as far as the so-called 'underprivileged' children are concerned. Such a conclusion would not in fact be altogether true; it is true only if we remain within a closed circle of thought which is governed by the notion that fundamentally all children have the same abilities, and that consequently all children should be taught by the same methods. We have seen that this is untrue even when we consider only personality; what is sauce for the introvert is most decidedly not sauce for the extrovert. It is even more true when we consider intelligence, and it is here that Jensen has made a most fruitful suggestion which deserves to be followed up to a much greater extent than has in fact been the case. We may begin by quoting some timely and relevant observations made by E. W. Gordon and D. A. Wilkerson, in their book on *Compensatory Education for the Disadvantaged*:

> . . . the unexpressed purpose of most compensatory programs is to make disadvantaged children as much as possible like the kinds of children with whom the school has been successful, and our standard of educational success is how well they approximate middle-class children in school performance. It is not at all clear that the concept of compensatory

education is the one which will most appropriately meet the problems of the disadvantaged. These children are *not* middle-class children, many of them never *will* be, and they can never be anything but second-rate as long as they are thought of as potentially middle-class children. . . . At best they are different, and an approach which views this difference merely as something to be overcome is probably doomed to failure.

In a similar vein, I have often pointed out that our educational system is geared to academic standards, and in particular to types of university education, which are inappropriate for the great majority of children submitted to this unholy educational mill which grinds exceeding small. Methods and aims which suit the academic child may not suit non-academic children, and to condemn them to competition and failure in a race in which they have no wish to take part in the first place is not a rational way to organize education.

As a reaction, many 'progressive' educationalists have gone to the other extreme. Seeing clearly that not all children can or should have academic types of education, they declare that none shall; educational objectives, instead of being geared to the most able academically, as in the past, are now geared to the least able academically, and even such far-reaching innovations as the abolition of examinations are seriously discussed. Underlying all these fads, past and present, is one fundamental error—that all children are alike, and must be given the same diet. Jensen's suggestion takes us away from this fundamental error. He begins (1973) with an observation often made by teachers of disadvantaged children:

many of these children seem much brighter than their IQs would lead one to expect, and . . . even though their scholastic performance is usually as poor as that of middle-class children of similar IQ, the disadvantaged children usually appear much brighter in non-scholastic ways than do their middle-class counterparts in IQ. A lower-class child coming into a new class, for example, will learn the names of 20 or 30 children in a few days, will quickly pick up the rules and the know-how of various games on the playground, and so on—a kind of performance that would seem to belie his IQ, which may be as low as 60. This gives the impression that the test is 'unfair' to the disadvantaged child, since middle-class children in this range of IQ will spend a year in a classroom without learning the names of more than a few classmates, and they seem almost as inept on the playground and in social interaction as they are in their academic work.

To explain these facts, Jensen proposes the existence of two genetically distinct basic processes underlying the continuum from simple associative learning to complex cognitive or conceptual learning. Ability thus appears at two distinct levels. Level I (associative ability) involves the neural registration and consolidation of stimulus inputs and the formation of associations. 'There is relatively little transformation of the input, so there is a high correspondence between the forms of the stimulus input and the form of the response output.' Level II (cognitive, conceptual ability) involves 'self-initiated elaboration and transformation of the stimulus input before it eventuates in an overt response. Concept learning and problem solving are good examples. The subject must actively manipulate the input to arrive at the output'. Intelligence tests, particularly tests of 'fluid' ability, are good measures of Level II ability: digit memory, rote learning, and free recall of visually or verbally presented materials (as in the scout test in which a tray full of objects is exposed for thirty seconds, and the scout has to repeat what he saw on it, after the tray has been covered by a sheet) are all good measure of Level I ability. These two types or levels of ability are not entirely uncorrelated, but the correlations are low enough to allow children of low IQ (poor Level II ability) to do very well on Level I ability tests. (The actual correlations are between .2 and .3, for samples of white, black and Mexican children.) Jensen finds that while differences in IQ (Level II abilities) between social classes are substantial, differences in Level I abilities are slight or non-existent; on the 'boy-scout test' mentioned above he found no difference between working-class and middle-class children, even though they differed some 15 to 20 points in IQ.

Jensen concludes his survey of the evidence by saying that:

> ordinary IQ tests are not seen as 'unfair' in the sense of yielding inaccurate or invalid measures for the many disadvantaged children who obtain low scores. If they are unfair, it is because they tap only one part of the total spectrum of mental abilities and do not reveal that aspect of mental ability which may be the disadvantaged child's strongest point—the ability for associative learning.

And he goes on to argue that:

> since traditional methods of classroom instruction were evolved in populations having a predominantly middle-class pattern of abilities, they put great emphasis on cognitive learning rather than associative learning. And in the post-Sputnik era, education has seen an increased emphasis on cognitive and conceptual learning, much to the disadvantage of many children whose mode of learning is predominantly associative. Many of the basic skills can be learned by various means, and an educational system that puts inordinate emphasis on only one mode or style of learning will obtain meagre results from the children who do not fit this pattern. At present, I believe that the educational system—even

as it falteringly attempts to help the disadvantaged
—operates in such a way as to maximize the importance of Level II . . . as a source of variance in
scholastic performance.

The recent stress on methods of 'discovery' is an excellent
example of this tendency; no conceivable method of
teaching could be better calculated to favour the middle-
class child and make learning difficult for the typical
working-class child! Discovery methods intentionally
disfavour associative learning, and stress the application
of Level II abilities, which are well developed only in
children with high g. It seems tragic that these methods
should be particularly popular in schools where the
children are least able to profit by them!

Basic skills, when taught in a manner dependent on
abstractive abilities, may not be learned at all by children low on g, and we often find such children in the
higher reaches of our schools who could easily have
learned these skills through associative methods, but
completely failed to learn them at all because they were
presented in a manner which did not take into account
the children's particular strengths and weaknesses. Jensen
continues:

> It may well be true that many children today are
> confronted in our schools with an educational philo-
> sophy and methodology which were mainly shaped
> in the past, entirely without any roots in these
> children's genetic and cultural heritage. The educa-
> tional system was never allowed to evolve in such a
> way as to maximize the actual potential for learning
> that is latent in these children's patterns of abilities.

Even worse, new methods (such as the 'discovery'
methods mentioned above) carry on and exaggerate the
faults of the older system—being developed by educationalists whose own high IQs make them unable to

understand the needs and different abilities of children with high Level I abilities, but deficient in Level II abilities.

Educational researchers must discover and devise teaching methods that capitalize on existing abilities for the acquisition of those basic skills which students will need in order to get good jobs when they leave school. I believe there will be greater rewards for all concerned if we further explore different types of ability and modes of learning, and seek to discover how these various abilities can serve the aims of education. This seems more promising than acting as though only one pattern of abilities, emphasizing *g*, can succeed educationally, and therefore trying to inculcate this one ability pattern in all children.

It might be thought that what is being recommended by Jensen is a sort of second-best education for some children, as compared with a superior type of education for others. This would be to look at his proposals in quite the wrong way. What Jensen is suggesting is simply that we should make use, in our educational venture, of the particular pattern of abilities which characterizes a child, or group of children, instead of forcing them to learn through the use of a pattern of abilities which, if not exactly lacking, is innately poorly developed in them. If a child of low IQ can learn to write, spell, do arithmetic, speak grammatically and generally acquire the basic skills which are needed by him or her to earn a reasonable living and lead a full and happy life, through the use of his associative abilities, it does not seem an act of kindness to force him to try to learn these skills through the use of conceptual abilities which he does not possess in adequate measure, and fail in the attempt. But that is what our schools are doing at the moment, all in the name of equality! To quote Jensen again:

The ideal of equality of educational opportunity should not be interpreted as uniformity of facilities, instructional techniques, and education aims for all children. Diversity rather than uniformity of approaches and aims would seem to be the key to making education rewarding for children of different patterns of ability. The reality of individual differences thus need not mean educational rewards for some children and frustration and defeat for others.

It seems clear to all who can see that our present educational policies are failing most comprehensively to engage the interest and the co-operation of that large group of children we call 'deprived' or 'underprivileged'. It would not be possible to guarantee that the use of associative methods of teaching would engage the active participation of these children, but it would seem worth a try. What may have soured their enthusiasm may have been their continuous failure to reach any satisfactory level of achievement through the use of current conceptual methods of teaching. Once they could be brought to see that success and achievement are possible for them, through the use of methods of teaching more closely geared to their particular pattern of abilities, this success and this achievement might be powerful incentives for increasing their interest and co-operation. Nothing succeeds like success; children who opt out of school have had a continued record of failure, and it would be difficult to blame the children themselves for voting with their feet and playing truant as much as possible. This failure is not necessary; it is imposed on the children by inappropriate methods of teaching which do not take into account the innate patterns of abilities of these children. A return to sanity is long overdue; we must pay close attention to the genetic basis of our children's abilities.

One reason why many people seem nowadays to despise associative learning seems to link up with the emphasis on 'creativity' already referred to in an earlier chapter, that is with the erroneous notion that tests of divergent ability define a different and superior aptitude to tests of convergent ability. Facts, and the knowledge of facts, are looked down upon as fit only for peasants and other people lacking in 'originality' and 'creativity', and the statement is repeated with approval that you do not need to know facts as long as you know where to look them up. Liam Hudson's little book on *The Cult of the Fact* (1972) seems to give expression to some such view, and there are many even more explicit statements to be found in the educational literature. This whole trend is dangerous. If my secretary cannot spell, it is no consolation to me that she can always look up words in her dictionary; in the first place she would not know which words she spelled wrongly, and would consequently fail to look them up, and in the second place drafts of manuscripts would never get finished if she had to take recourse to the dictionary at all times. Similarly, a historian must have a large scaffolding of dates in order to place new information; he can of course look up any particular date, but without such a scaffolding he simply cannot place events and relations between events in any meaningful context. For a scientist, particularly an innovative or 'creative' one, a wide knowledge of all the relevant facts is an absolute must; without it he would not even know whether his thoughts were or were not original! Edison said that genius was 90% perspiration and 10% inspiration; the perspiration derives from the effort to get to know all the relevant facts (and all the facts which might be relevant, but turn out not to be) and without it the inspiration simply cannot arise, or else is still-born. Modern education does no favour to the children it is supposed to teach when it de-emphasizes facts; although facts are not the only important things in life, in science, and in the arts, they nevertheless constitute the absolutely essential substructure without which nothing worthwhile can be built.

9

Even when we have learned to live with the idea that all modification of behaviour depends on the interaction between genetic and environmental factors, we still have to face quite complex problems in deciding upon social action. The precise nature of these problems is seldom verbalized, and even more seldom is it put into a proper quantitative form. I will first of all discuss a small-scale empirical study which will serve to illustrate the sort of thing that can and should be done, and then go on to discuss some broader applications of the notions introduced. The work to be considered first is that of R. C. Atkinson (1972), of Stanford University; it is concerned with the development of a computer-assisted instruction programme for teaching reading in the primary grades (CAI for short). This programme provides individualized instruction in reading and is used as a supplement to normal classroom teaching; a given student may spend anything from o to 30 minutes per day at a CAI terminal. Performance is measured on a standardized reading achievement test. Now it is possible to construct a statistical model which predicts the child's test performance as a function of the time he spends on the CAI system.* The next step is to decide how we can best

---

* If we let $P_i(t)$ be student i's performance on a reading test administered at the end of first grade, given that he spends time t on the CAI system during the school year, then within certain limits, the following equation holds:

$$P_i(t) = \alpha_i - \beta_i \exp(-\gamma_i t)$$

Depending on a student's particular parameter values, the more time spent on the CAI program, the higher the level of achievement at the end of the year. The parameters α, β and γ characterize a given student and vary from one student to the next; α and $(\alpha - \beta)$ are measures of the student's maximal and minimal levels of achievement, respectively, and γ is a rate of progress measure. These parameters can be estimated from a student's response record obtained during his first hour of CAI. Stated otherwise, data from the first hour of CAI can be used to estimate the parameters α, β, and γ for a given student, and then the above equation enables one to predict

allocate scarce resources (time on the computer) to the children in such a way as to obtain the best results. But how are we to decide which results are 'best'? Normally such decisions are made without proper consideration of the alternatives, on the basis of political or social prejudice, and without knowledge of the actual outcomes involved. In the present case, however, the formula (which is reasonably accurate—sufficiently so in any case to give a quantitative answer to our questions) enables us to make quite explicit statements about the outcome of different actions. Let us look at four different objectives that might be chosen by the headmaster of the school, or by a council of parents, or by the government (P stands for performance). These objectives are:

(a) Maximize the mean value of P over the class of students.

(b) Minimize the variance of P over the class of students.

(c) Maximize the number of students who score at grade level at the end of the first year.

(d) Maximize the mean value of P satisfying the constraint that the resulting variance of P is less than or equal to the variance that would have been obtained if no CAI were administered.

Objective *a* maximizes the gain for the class as a

---

end-of-year performance as a function of the CAI time allocated to that student.

We are now faced with a rather interesting problem. Computer time is expensive. Suppose that a school has budgeted a fixed amount of time T on the CAI system for the school year and must decide how to allocate the time among a class of n first-grade students. Assume, further, that all students have had a preliminary run on the CAI system so that estimates of the parameters α, β, and γ have been obtained for each student. Let t_i be the time allocated to student i. Then the goal is to select a vector $(t_1, t_2, \ldots, t_n)$ which optimizes learning.

whole; Objective *b* aims to reduce differences among students by making the class as homogeneous as possible; Objective *c* is concerned specifically with those students who fall behind grade level; Objective *d* attempts to maximize performance of the whole class but insures that differences among students are not amplified by CAI.

If we follow objective *(a)* we find that the mean performance of the class is 15% higher than if we had allocated time equally to all children; this is a sizeable gain. Unfortunately there is also an increase in variance of roughly 15%; in other words, the difference between the best readers and the worst readers has increased! If we rather accept objective *(b)* we find that compared with an equal distribution of time at the terminal, there is a reduction of overall performance of 15%, but there is also a reduction of 12% in variability. Adopting policy *(c)*, we have a reduction in overall performance, compared with equal time allocation, of 9%, and also a reduction in variability of 10%. Objective *(d)*, which attempts to strike a balance between *(a)* on the one hand, and *(b)* and *(c)* on the other, yields an increase in performance, over equal time sharing, of 8%, and yet reduces variability by 6%. Objective *(a)* is the most 'meritocratic', objectives *(b)* and *(c)* are the most 'mediocratic', while objective *(d)* attempts to maximize both the desiderata that there should be maximum improvement, but also a reduction in variability. It is by spelling out the consequences of social action in this precise, numerical fashion that we can approach the giving of rational answers. Politicians would more usually have approached the problem by mouthing platitudes about excellence or equality, but they would not have been able to offer a precise and meaningful choice between objectives. Our whole educational establishment is devoted to the purely verbal approach; until a more quantitative approach becomes fashionable our decisions on important social objectives will not be made on rational grounds.* Human diversity has to be quantified, and inserted in proper prediction equations before rational decisions can be made.

Most attempts to deal with complex educational and other social issues along political lines hopelessly over-simplify the case, apart from failing to introduce the needed quantitative estimates of all parameters without which no reasonable conclusion can be arrived at. Consider such an issue as comprehensive schools, and the abolition of the grammar and public schools. To most people the argument is presented in some such terms as: 'Are comprehensive schools a good thing or a bad thing?', and the only alternative answers permitted are 'Yes, definitely' or 'No, certainly not'. Similarly, the policy of 'streaming' within comprehensive schools is attacked and defended along equally simplistic lines. Yet clearly such questions and answers are completely meaningless. There is nothing corresponding to the term 'comprehensive school', or 'streaming'; there are many different types of comprehensive schools, some good, some bad, and there are many different systems of streaming, some good, some bad. Without detailed specification of the exact type of comprehensive school, or of method of streaming, only the most extreme bigot would have a ready answer to the kind of question asked above.

Even when detailed specifications are given, the question is still too broad. The terms 'good' and 'bad' are still left undefined; do they refer to scholastic achievement, or to social integration, or what? It is quite possible that a given policy may lead to greater social

* Parameters such as α, β and γ in the equation are defined in terms of actual performance on the test, but they are undoubtedly closely linked with genetically determined factors, such as intelligence, and possibly personality; similar formulae for wider-ranging educational objectives and methods would certainly have to include such parameters.

cohesiveness, but to a lowering of scholastic achievement; is such a policy good or bad? Can we answer the question at all without having some quantitative estimate of just how much a policy affects scholastic achievement and how much it affects social cohesiveness? No rational decisions are possible unless we have at least some estimates of the sort of magnitude of effect we can expect; without that we are deciding on purely irrational, emotional grounds. Indeed, the expected effects might not occur at all, or it is even possible that the opposite effects might be found; we have already seen that the introduction of 'quota' systems in American schools, designed to lead to greater integration and social cohesiveness, have led instead to poorer integration and less social cohesiveness. 'Gut reactions' are not a reliable guide to social action, particularly when these 'gut reactions' are based on suspect premises, prejudice, and ignorance of vital facts.

Even this more adequate way of looking at the problem is far too simplistic; it assumes that different children will react in a similar manner to certain changes in educational policy. Our knowledge of genetic differences in ability and personality makes this extremely unlikely. It seems to make sense to many educationists to ask questions such as: 'Is programmed instruction via computer better than old-fashioned teaching?' but of course such a question is nonsense, even if we agreed on the precise method of measuring the effects in question. There are good and bad programmes, good and bad teachers; a good programme might be better than a bad teacher, a good teacher better than a bad programme. But even more important, high-ability and low-ability children may react quite differently to teaching programmes; some of the constants in Atkinson's equation refer to ability parameters. It is also known that personality plays an important part in deciding how an individual child will react to computer teaching; introverts have been found to react well, liking the impersonal method of learning, while extroverts do better with the old-fashioned, personalized method.

Similar findings are available for such new concepts in teaching as the 'discovery' method; extroverts like it and do well with it, introverts do not. It is customary to discuss these problems as if all children were identical twins, reacting uniformly to different methods of teaching, but the genetically determined differences between people make such a conception quite incapable of dealing with the complexities of real life. Different people react quite differently to given stimuli or sets of stimuli; recognition of this fact is the beginning of wisdom in psychology, in education, and in social life generally. Social reformers who fail to realize this fact are likely to make things worse, rather than better.

Should they be blamed for this? Perhaps only if they are extroverts. There are several experimental studies (including one from behind the Iron Curtain) which show that introverts react positively to praise, negatively to blame, while extroverts reverse this pattern. These studies deal with school children, but perhaps the conclusion is valid for others too.

These principles have been applied to the problem of 'streaming' in a very interesting unpublished study by Dr Holden. He compared the scholastic achievement and the social competence and happiness of children under conditions of streaming and no streaming, using different classes as their own controls, that is studying the same class under both conditions. He also tested the intelligence and personality of the children involved, and in addition selected teachers who either approved or did not approve of streaming. His results are far too complex to make it possible to summarize them here; what did emerge clearly from his work was that no general conclusions were possible—in other words, no conclusions were true equally for introverts and extroverts, for teachers favourable and unfavourable to streaming, for bright and dull children, or for educational and social objectives. There is a bewildering, but not meaningless, combination of significant effects; the very idea of deriving a single answer in terms of 'streaming is good', or 'streaming is bad' from these data is too nonsensical to be even contemplated.

Work on the effects of ability grouping by Goldberg, Passow and Justman (1966) shows just how difficult it is to make reasonable generalizations in this field. It is intuitively obvious that teachers will find their task easier, and children will progress better, when the range of ability in a particular class is narrow, rather than so wide as to include very bright and very dull children. Yet this has not always been found to be so: 'narrowing the range of ability (on the basis of group intelligence tests) *per se*, without specifically designed variations in program for the several ability levels, does not result in consistently greater academic achievement for any group of pupils.' Yet many studies did find greater achievement for special classes of bright children. 'Those studies which have found advantages in narrowing the range . . . have usually dealt with programs in which groups of gifted pupils were either accelerated through the standard sequence or were exposed to content not normally taught. In both cases, pupils not only "did as well" as others on the standard achievement tests covering the expected grade-level content, but also showed considerable knowledge of the additional material.' Narrowing the range of ability must be accompanied by well-designed changes in syllabus, method of teaching, and comprehensiveness of instruction, in order to produce the expected benefits; teachers have to be specially trained to put these changes into effect. There are no simple solutions in this field, and even reasonable expectations are not always fulfilled unless steps are taken to utilize administrative actions to their best advantage.

Yet educational research is still enmeshed in busy work which does not take into account any of the complexities pointed up in studies of this kind, and politicians still talk as if their policies had any general validity. Until this complex kind of research is done on a large enough scale to give us a glimpse at least of general, scientifically valid laws and interactions, we are not likely to achieve any of our objectives in education, however many arbitrary 'solutions' we impose on our defenceless children. Human beings are diverse; they are born diverse and life experiences serve if anything to make them more diverse yet. It is not reasonable to treat them as if this diversity did not exist, and while such disregard certainly makes political advocacy easier, it does not help in achieving such aims as we may formulate for our society.

These considerations also apply to quite a different field than that of education: indeed, it would be difficult to name any social field whatever to which they would not apply. Consider those two ugly sisters of modern Marxist sociology, *anomie* and *alienation*. We are assured, on the basis of oddly little evidence, that modern society, and those living in it, are having their lives blighted by these reaction patterns to the stresses of modern industry. Hag-ridden by conveyor-belt methods of production, 'we have no time to stand and stare'; the result is unhappiness, distress, and general discomfort, as well as an abiding dislike of industrial production. Oddly enough, whenever polling organizations question modern man about the degree of contentment he shows in his work, the figures demonstrate exactly the opposite, namely considerable contentment and indeed liking for whatever his job may be.

The latest figures to appear were taken from such a poll in Austria; they showed that 41% of a random sample of the population were very contented, while another 49% were contented. 7% were neither contented nor discontented, while only 3% were discontented or very discontented. With something like 90% contented, advocates of the notion of anomie and alienation would seem to have some difficulty in proving their case; at most it would seem to apply to the 3% who were discontented in this sample! Even if the figures were much higher in other countries (and the evidence suggests that they are not), they would still show much more contentment than discontentment.* Why is it that these ideas have attracted so much support among middle-class writers?

The most obvious reason of course is that many people judge the attractiveness of someone else's job in terms of their own likes and dislikes. I would not like to be a politician, a worker at an assembly line, a salesman, a miner, or an actor—for various reasons; if I am not careful I may go on to argue that politicians, workers at assembly lines, salesmen, miners, or actors are unhappy in their jobs, and suffering from anomie and alienation. But conversely, I am sure that politicians, salesmen, miners, actors, etcetera, would hate to have my job, and might equally argue that I must be suffering from anomie and alienation! It is one of the most beneficent aspects of human diversity that different people have different abilities, different personalities, different temperaments, which suit them to different types of jobs. Where the introvert would be happy with such a job as, say, a gardener's, where he could work in isolation, an extrovert would be happy with such a job as, say, a salesman's, where he could make use of his sociable tendencies. Give the wrong job to either, and we might indeed have trouble on our hands!

Even such jobs as working on a car assembly line, often described as 'soul destroying' by writers whose personality might indeed be such as to make them dislike such a job were it offered to them, are in fact liked

* In a recent French poll, 89% of those questioned said that they were either fairly happy or very happy. In England, less than 1 in 8 of a large, representative sample of workers declared that he was not interested in his work, leaving 87% who were actively interested. American polls, too, show only about 10% at most 'alienated' and disaffected. These are interesting figures, difficult to reconcile with the Marxist hypothesis.

by many of the people 'on the line'; just as the dangerous, hard and dirty job done by the miner is by no means disliked as much as one might think at first sight. Perhaps the fact that much of the miner's work is done in small groups, and that he can largely determine for himself just how to organize his work, makes the difficulties acceptable to one kind of personality; perhaps the fact that the assembly line job protects the worker from interruption by other people and leaves him free to attend to his own thoughts, suits his particular personality.

Almost no work has been done to determine whether those who are least contented with their respective jobs (from the point of view of job content, not of pay) may not be those temperamentally least suited to those particular jobs. Vocational guidance and occupational selection have concentrated far more on a person's abilities than on his personality; even so, there is considerable evidence to show that many jobs have a personality profile conformity which determines to a surprising degree whether or not a person will be happy and contented (as well as successful) in a particular job.

As a demonstration of the degree to which success at a given job may be related to personality, consider a study already alluded to in Chapter 5. Götz and Götz (1973) set out to test the hypothesis that gifted artists in the visual sphere (painting) would predominantly be emotional introverts, that is have low scores on E and high scores on N. Figure 13 shows the outcome of testing fifty ungifted, thirty-five gifted, and fifteen highly gifted students working in separate studios in an art academy. Ratings were made by art experts who had worked with these students for periods of four to six years. These ratings were made, not only of technical skill, but also of originality, independence, self-actualization, achievement motivation, and aptitude for fantasy. Figure 13 shows the scores for the fifty ungifted students (B), the thirty-five gifted students (A), and individually for the

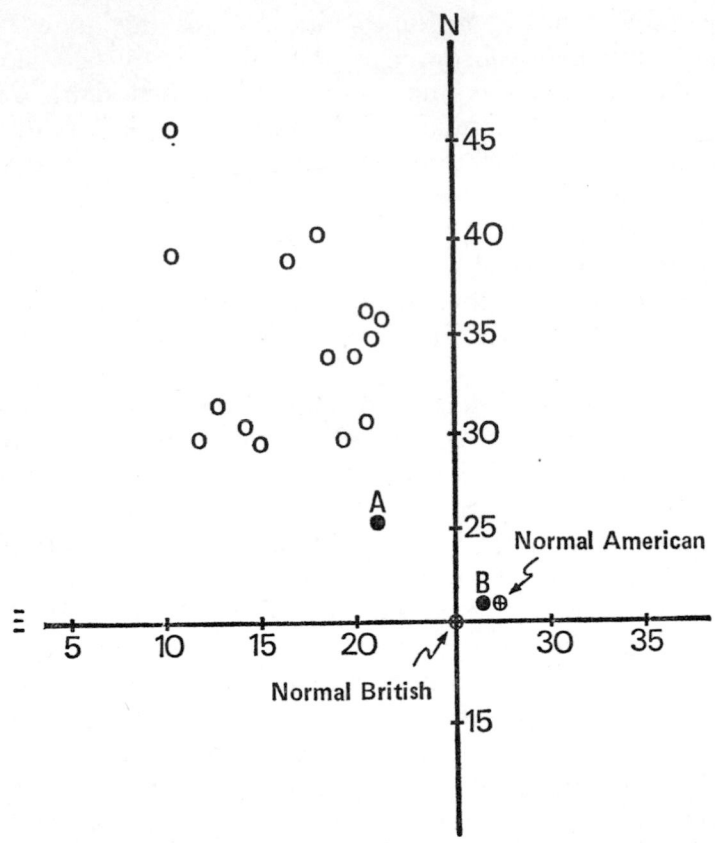

Figure 13   *The scores of 15 highly gifted, 35 gifted and 50 ungifted students*

fifteen highly gifted students. Note that the ungifted are very close to the mean of the British and American standardization samples; the gifted are significantly more introverted and higher on N than the ungifted. But note also the individual scores of the fifteen highly gifted; each one of them is found in the low E-high N quadrant, and each one of them shows the predicted temperamental traits to a more marked extent than the average 'gifted' student! Clearly, there is a surprisingly close relationship between personality and achievement in this field—a field, in addition, where it is very difficult indeed to quantify the factors entering into success.

Consider now the personality type which we would expect to find in the quadrant opposite to our successful, highly gifted artists, that is the stable extrovert. As mentioned in Chapter 5 studies have been done by various army groups of the personality of the successful commando and paratrooper, and the outcome was precisely the mirror image of that described in Figure 13; almost all commandos and paratroopers fell into the stable extrovert quadrant! There was practically no overlap between successful, highly gifted artists, and successful, highly commended warriors. There is of course a natural selection all along the line before a person becomes successful in either of these two widely different lines of endeavour; this selection process ensures that people not adapted to a particular type of work do not in fact ever get as far as being accepted into it.

This selection process seems to work reasonably well in the examples here considered; it probably does not work at all well in most jobs which the man in the street is likely to consider in his youth. There is of course much to be said for the 'job enrichment' practices which are at the moment gaining in popularity, and nothing said here should be taken to argue against the need for such innovations. Nevertheless, it would seem to me that much could also be gained, as far as job contentment is concerned, if we took seriously the facts of human diversity, and tried to put round pegs into round holes, not only with respect to ability, but also with respect to personality. Much of the work that man has to do is hard, difficult, and repetitive; but there is no reason why it should not also be rewarding and enjoyable. Selection procedures which ensure as far as possible that man and work should be suited to each other may do much to make life more pleasant and happier all round; the diversity of human abilities and temperamental traits ensures that such selection will be a great improvement on the largely random method of choice so prevalent at the moment.

It will now be clear why I regard the message of the facts recounted in this book as being of the utmost importance for society, and why I believe that the social consequences which these facts may have are vitally relevant to our attempts to extricate ourselves from the evils of past days, and to improve the lot of humankind everywhere. Reform of what is wrong in our society there must be, but unless this reform takes into account limitations set by inexorable biological facts it is likely to achieve nothing. Several hundred years of experience with physics and chemistry has taught us that we must co-operate with nature; we cannot coerce her. The same is true of psychology. We must learn to co-operate with nature; attempts to disregard her laws and get our way against her opposition are doomed to failure. Science and technology have been responsible for raising us from the level of beasts of burden to our present uncertain eminence; unless we learn to apply methods of scientific reasoning to our own behaviour we are not likely to remain the cynosure of nature's regard for long. Where hopes are dupes, fears may by liars; nevertheless, our present position is too fraught with danger for us to be complacent. Recognition of man's biological nature, and the genetically determined inequality inevitably associated with his derivation, is an absolutely necessary beginning for any attempt to use the methods of science and reason in an effort to save ourselves from the very real dangers that confront us.

References, Acknowledgements, Index

References

Atkinson, R. C., 'Ingredients for a theory of instruction', *Amer. Psychologist*, 1972, 921–31.

Binet, A. & Simon, T., 'Méthodes nouvelles pour le diagnostic du niveau intellectuel des anormaux', *Année Psychol.*, 1905, *11*, 191–277.

Bloom, B. S., *Stability and change in human characteristics*, Wiley, 1965.

Burks, B. S., 'The relative influence of nature and nurture upon mental development', *Yearbook of the National Society for Studies in Education*, 1928, *27(I)*, 219–316.

Burt, C., 'The genetic determination of differences in intelligence: a study of monozygotic twins reared together and apart', *Brit. J. Psychol.*, 1966, *57*, 137–53.

'Class differences in general intelligence: III', *Brit. J. Statl. Psychol.*, 1959, *12*, 15–23.

'Intelligence and social mobility', *Brit. J. Statl. Psychol.*, 1961, *14*, 3–24.

Cattell, R. B., 'Theory of fluid and crystallized intelligence: a critical experiment, *J. Educ. Psychol.*, 1963, *54*, 1–22.

Abilities, their structures, growth and action, Houghton Mifflin, 1971.

Coleman, J. S. *et al*, *Equality of educational opportunity*, Washington, DC, US Office of Education, 1966.

Cronbach, L. & Meehl, P. E., 'Construct validity in psychological tests', *Psychol. Bull.*, 1955, *52*, 281–302.

Crowe, R. R., 'The adopted offspring of women criminal offenders: a study of their arrest records', *Arch. Gen. Psychiat.*, 1972, *27*, 600–03.

Darlington, C. D., 'Heredity and Environment', *Proc. IX Internat. Congr. Genetics, Caryologia*, 1954, 190.

Davie, R., Butler, N., & Goldstein, H., *From birth to seven*, Longman, 1972.

Dockrell, W. B. (ed.), *On intelligence*, Methuen, 1970.

Erlenmeyer-Kimling, L. & Jarvik, L. F., 'Genetics and intelligence: A review', *Science*, 1963, *142*, 177–9.

Ertl, J. P., & Schafer, E. W. P., 'Brain response correlates of psychometric intelligence', *Nature*, 1969, *223*, 421–22.

Eysenck, H. J., *The biological basis of personality*, C. C. Thomas, 1967.

Crime and personality (2nd edn.), Paladin, 1970.

Race, intelligence and education, Temple Smith, 1971.

Psychology is about people, Allen Lane, 1972.

'An experimental and genetic model of Schizophrenia', in A. R. Kaplan (ed.), *Genetic Factors in 'Schizophrenia'*, C. C. Thomas, 1972.

The measurement of intelligence, Medical and Technical Publishing Co., Lancaster, 1973.

Floud, J. E., & Halsey, S. H., *Social class and educational opportunity*, Heinemann, 1956.

Getzels, J. W., & Jackson, P. W., *Creativity and intelligence*, Wiley, 1962.

Götz, K. O. & K., 'Introversion-extroversion and neuroticism in gifted and ungifted art students', *Perceptual and Motor Skills*, 1973, *36*, 675–78.

Glass, D. C., *Genetics*, Rockefeller University Press, 1968.

Glass, D. V., *Social mobility in Britain*, Routledge & Kegan Paul, 1954.

Goldberg, M. C., Passow, A. H., & Justman, J., *The effects of ability grouping*, Teachers College Press, 1966.

Gordon, E. W. & Wilkerson, D. W., *Compensatory Education for the Disadvantaged*, College Entrance Examination Board, New York, 1966.

Guilford, J. P., *The nature of human intelligence*, McGraw-Hill, 1967.

& Hoepfner, R., *The analysis of intelligence*, McGraw-Hill, 1971.

Herrnstein, R., *IQ in meritocracy*, Allen Lane, 1973.

Hirsch, J., *Behaviour-genetic analysis*, McGraw-Hill, 1967.

Horn, J. L., & Knapp, J. R., 'On the subjective character of the empirical base of the structure-of-intellect model', *Psychol. Bull.*, 1974, to appear.

Hudson, L., *The cult of the fact*, Jonathan Cape, 1972.

Hughes, K. R., & Zubek, J. P., 'Effects of glutamic acid on the learning ability of bright and dull rats', *Canadian J. Psychol.*, 1956, *10*, 132–38.

Hull, C. L., *Principles and behaviour*, Appleton-Century, 1943.

Hutt, Corinne, *Males and females*, Penguin, 1972.

Jensen, A. R., 'Environment, heredity and intelligence', *Harvard Educ. Rev.*, Reprint Series No. 2, 1969.
Genetics and education, Methuen, 1972.
Educability and group differences, Methuen, 1973.

Jinks, J. L., & Fulker, D. V., 'Comparisons of the biometrical and genetical, MAVA and classical approaches to the analysis of human behaviour, *Psychol. Bull.*, 1970, *73*, 311–349.

Joffe, J. M., *Prenatal determinants of behaviour*, Pergamon Press, 1969.

Johnson, R. C., 'Similarity in IQ of separate identical twins as related to length of time spent in same environment', *Child Development*, 1963, *34*, 745–49.

Lakatos, I., & Musgrave, A., *Criticism and the growth of knowledge*, CUP., 1970.

Lange, J., *Verbrechen als Schicksal* (Crime as Destiny), Thieme, Leipzig, 1929.

Lawrence, E. M., 'An investigation into the relation between intelligence and inheritance', *Brit. J. Psychol.*, Monograph Supplement, 1931, *16*, No. 5.

Leahy, A. M., 'Nature-nurture and intelligence', *Geretics Psychol. Monographs*, 1935, *17*, 241–305.

Li, C. C., 'A tale of two thermos bottles: properties of a genetic model for human intelligence', in R. Cancro (Ed.), *Intelligence: genetic and environmental influences*, Grune & Stratton, 1971, pp. 162–81.

Liddicoat, R., 'The effects of maternal antenatal decompression treatment on infant mental development'. *Psychologia Africana*, 1968, *12*, 103–21.

Lipset, S. M., & Bendix, R., *Social mobility in industrial society*, Heinemann, 1959.

Middleton, W. E. Knowles, *A history of the thermometer*, Johns Hopkins Press, 1966.

Newman, H. H., Freeman, F. N., & Holzinger, K. J., *Twins: a study of heredity and environment*, Univ. Chicago Press, 1937.

Page, E. B., 'Miracle in Milwaukee: raising the IQ', *Educational Researcher*, 1972, *1*, 8–16.

Pascal, B., 'Récit de la grande expérience de l'équilibre des liqueurs', *Oeuvres*, 1649.

Piaget, J., *The psychology of intelligence*, Harcourt, Brace & World, 1950.

Price, B., 'Primary biases in twin studies', *Amer. J. Human Genetics*, 1950, 2, 293–355.

Rachman, S., 'An examination of Laing's views on schizophrenia', *New Society*, 26 April 1973.

Reed, Elizabeth, & S. C., *Mental retardation*, Saunders, 1965.

Rosenthal, R., & Jacobson, L., *Pygmalion in the classroom*, Holt, Rinehart & Winston, 1968.

Schull, W. J., & Neel, J. V., *The effects of inbreeding in Japanese children*, Harper & Row, 1965.

Schulsinger, F., *Psychopathy; heredity and environment*, *Int. J. ment. Health*, 1972, *1*, 190–206.

Shulz, B., Quoted by Shields, 1973. Article quoted appeared in 1951.

Shields, J., *Monozygotic twins brought up apart and brought up together*, OUP, 1962.
   'Heredity and psychological abnormality' in H. J. Eysenck (Ed.), *Handbook of Abnormal Psychology*, 2nd edn., Pitman, 1973.

Snow, R. E., 'Unfinished Pygmalion', *Contemporary Psychology*, 1969, *14*, 197–99.

Spearman, C., *The abilities of Man*, Macmillan, 1927.

Stein, Z., Susser, M., Saenger, G., & Marolla, F., 'Nutrition and mental performance'. *Science*, 1972, *178*, 708–13.

Terman, L. M., & Oden, M. N., *The gifted groups at mid-life*, Stanford University Press, 1959.

Thoday, J. M., 'Geneticism and environmentalism', in J. E. Meade and A. S. Parkes (Eds.) *Biological Aspects of Social Problems*, Oliver & Boyd, 1965.

Thorndike, E. L., *The measurement of intelligence*, Columbia Univ., 1927.

Thurstone, L. L., *Primary mental abilities*, Univ. Chicago Press, 1938.

Thurstone, L. L., & T. G., *Factorial studies of intelligence*, Univ. Chicago Press, 1941.

Tyler, L. E., *The psychology of human differences*, Appleton-Century-Crofts, 1956.

Tuddenham, R. D., 'A "Piagetian" test of cognitive development', in B. Dockrell (Ed.), *On Intelligence*, Methuen, 1970.

Vernon, P. E., 'Ability factors and environmental influences', *Amer. Psychologist*, 1965, *20*, 723–33.

'Intelligence' in B. Dockrell (Ed.), *On Intelligence*, Methuen, 1970.

Wallach, M., & Kogan, N., *Modes of thinking in young children*, Holt, Rinehart & Winston, 1965.

Waller, J. H., 'Achievement and social mobility', *Social Biology*, 1971, *18*, 252–59.

Wilson, M., *Passion to know*, Weidenfeld, 1972.

Wilson, R. S., 'Twins: early mental development', *Science*, 1972, *175*, 914–17.

Acknowledgements

The author gratefully acknowledges permission to quote the following copyrighted material in this book:
J. B. S. Haldane, *The Inequality of Man*, Chatto & Windus, 1932, and Mrs Helen Spurway Haldane; E. B. Page, 'Miracle in Milwaukee: Raising the IQ', *Educational Researcher*, 1972, *1*; C. C. Li, 'A tale of two thermos bottles' in *Intelligence: genetic and environmental influences*, R. Cancro (ed.), Grune & Stratton, 1973; C. Hutt, *Males and Females*, Penguin Books, 1972; R. Herrnstein, *IQ in the meritocracy*, Allen Lane, 1973, first published as 'IQ' in *Atlantic Monthly*, September 1971; A. Jensen, *Educability and Group Differences*, Methuen & Co, 1973; R. C. Atkinson, 'Ingredients for a theory of instruction', *American Psychologist*, 1972.

For the problems on pages 47-49: H. J. Eysenck, *Check your own IQ*, Penguin Books, 1966.

For the figures: (6) A. Jensen, *Genetics and Education*, Methuen & Co, 1972; (7 & 8) C. C. Li, 'A tale of two thermos bottles' in *Intelligence: genetic and environmental influences*, Grune & Stratton, 1973; (9) J. Waller, *Social Biology*, 1971, 18, 252-9; (13) K. O. & K. Gotz, 'Introversion – extroversion and neuroticism in gifted and ungifted art students', *Perceptual and Motor Skills*, 1973, *36*, 677.

For the tables: (2) L. L. & T. C. Thurstone, *Factorial studies in intelligence*, Chicago Univ. Press, 1941; (5, 8 and 9) C. Burt, 'Intelligence and social mobility', *Brit. J. Statl. Phsychol*, 1961, 14; (7) C. Burt, 'Class differences in general intelligence', *Brit. J. Statl. Psychol.*, 1959, *12*; (10) B. S. Bloom, *Stability and change in human characteristics*, Wiley, 1965; (13) R. Liddicoat, 'The effects of antenatal decompression treatment', *Psychologia Africana*, 1968, *12*.

Index